PRESIDENTIAL TAKEDOWN

PRESIDENTIAL TAKEDOWN

*How Anthony Fauci, the CDC,
NIH, and the WHO Conspired to
Overthrow President Trump*

DR. PAUL ELIAS ALEXANDER
AND KENT HECKENLIVELY, JD

Skyhorse Publishing

Skyhorse Publishing books may be purchased in bulk at special discounts for sales promotion, corporate gifts, fund-raising, or educational purposes. Special editions can also be created to specifications. For details, contact the Special Sales Department, Skyhorse Publishing, 307 West 36th Street, 11th Floor, New York, NY 10018 or info@skyhorsepublishing.com.

Skyhorse® and Skyhorse Publishing® are registered trademarks of Skyhorse Publishing, Inc.®, a Delaware corporation.

Visit our website at www.skyhorsepublishing.com.

10 9 8 7 6 5 4 3 2 1

Library of Congress Cataloging-in-Publication Data is available on file.

Cover image from Wikimedia Commons

Print ISBN: 978-1-5107-7622-7
Ebook ISBN: 978-1-5107-7623-4

Printed in the United States of America

The party told you to reject the evidence of your eyes and ears. It was their final, most essential command.

—George Orwell, 1984

Contents

Schooling Dr. Fauci

It's difficult to overstate how badly the United States government is being run.

I should know.

I am Paul Elias Alexander, PhD, brought into the US government in May 2020 as their "evidence-based" medical expert to ensure that the government's response to the COVID-19 crisis was based on solid science. My principal role was to advise senior government officials working out of the Health and Human Services (HHS) building in Washington, DC, as to the COVID pandemic response. Tasks would be varied, and my title was that of "Senior COVID Pandemic Advisor."

I must state emphatically that the government's response was not based on science, and I had little hesitation in making my thoughts known. Officials stated that they were "following the science," yet it became clear to me that they were actually averse to the science and disregarding it entirely. It was as if the real science and data did not exist.

Although I'm a Canadian citizen, I'd been lured into working for the United States by the siren call of being able to apply my special expertise to perhaps the greatest historic challenge to public health, only to find myself chewed up and spit out. Now more than two years after I accepted the position, I struggle to make a stable income. I was told very openly that because I worked for the Trump administration and because I raised questions as to the lockdown and school closure response, including the failure to use early drug treatment, then I should consider my career in DC, and in general, over.

If anything, the COVID-19 crisis was when we should have seen government officials acting at their very best and in the public interest.

They did not.

Let me give you a glimpse into the "belly of the beast" at a high-level meeting and exchange of communications, where I argued that evidence showed that schools should be reopened, as President Trump had been advocating.

But evidence was not what they wanted; only fear and continuing lockdowns. And although I didn't know it at the time, my fate had already been decided.

* * *

It's important to understand my position at the Department of Health and Human Services (DHHS), where I worked from approximately May 10, 2020 to September 25, 2020 when I was forced to resign. I had the White House on one phone telling me I needed to stay and a human resources person from HHS shouting on a separate phone that they controlled Washington, DC, not the president.

When I arrived in Washington, DC, I was assigned to work with Mr. Michael Caputo, then the assistant secretary of Health and Human Services for Public Affairs. Given my background in evidence-based medicine, having done master's work at York University, Canada, having completed a master's degree in epidemiology from the University of Toronto, a master's in evidence-based medicine from Oxford University, and a doctorate in evidence-based medicine and research methods from McMaster University in Canada, I was hired in May 2020 to be the COVID pandemic advisor in HHS, and would function as the liaison between HHS and the White House, among many other duties. Just prior to going to Washington, DC, and for the first three weeks while in the Trump administration, I also worked for the World Health Organization (WHO) in Geneva, Switzerland and Pan American Health Organization (PAHO) out of Washington, DC, as a COVID pandemic evidence synthesizer.

I also think it's vital for my readers to understand that McMaster University, although not a familiar name to most Americans, is considered the world's leader in evidence-based medicine (EBM), ahead of even Stanford or Harvard Universities. The field of EBM was founded by my doctoral supervisor, Dr. Gordon Guyatt, in Hamilton, Ontario, at McMaster University. In addition to my former position as an assistant professor of

EBM and research methodology at McMaster, I've worked for the WHO as a regional specialist/epidemiologist for Europe's Regional Denmark office with responsibility for Russia, Turkey, Ukraine, and Poland behavioral risk projects (which were funded by Bloomberg L.P.). I also worked for twelve years for the Canadian government at Health Canada (Population and Public Health Branch), two of those years as the in-field epidemiologist for South Asia, covering India, Pakistan, Nepal, Sri Lanka, Bangladesh, Bhutan, the Maldives Islands, and Afghanistan for tuberculosis (TB) and HIV control (including drug resistant TB). During those years, from 2002 to 2004, I was posted in Kathmandu, Nepal, where on a cloudless day one might even glimpse the mighty Himalayas. From 2017 to 2019, I worked for the Infectious Diseases Society of America (IDSA) as their lead trainer in evidence synthesis and the development of systematic review guidelines.

In contrast to my extensive scientific background, my boss, Michael Caputo, was a longtime political strategist and lobbyist. He'd started his career working for Congressman Jack Kemp, and then learned a great deal more about politics from controversial figure Roger Stone when he became the lobbyist's personal driver. Among other jobs Caputo has held, he worked with Lt. Colonel Oliver North in the Reagan White House, then as director of media services for President George H. W. Bush in the 1992 campaign, and after that he served as an advisor to Russian President Boris Yeltsin.

During the 2016 campaign, Michael Caputo was in charge of communications in New York for the Trump campaign. Although Michael did not have a scientific background, I always found him to be highly intelligent, quick to grasp what I was telling him, well-connected, and a man with great integrity who wanted the best for America. It is important that I say this to you, the reader. Michael Caputo revealed himself, in my dealings with him, to be one of the most principled people I've met and someone who loves America and all of the good things America stands for. It was that which allowed me to have such a good professional relationship with him. I have nothing but praise for him.

In early September 2020, Trump's re-election campaign was in full swing and he had just finished a three-day bus trip through Texas.[1] The big question on everybody's mind was whether schools might soon reopen. By this time, there had already been several dustups between Trump and members of the COVID-19 task force regarding the facts and proper course of action to be taken. Trump wanted the schools reopened. Trump wanted society reopened, and he constantly made that clear to the COVID task force, to the media, and to the public. To the extent that I may share and not violate confidentialities, I can tell you that President Trump had many heated discussions with the task force members and mainly Dr. Anthony Fauci and Dr. Deborah Birx about their reluctance and insistence that schools remain closed, that society remain closed, that businesses remain closed, and that mask mandates that included young children were necessary. These heated, and oftentimes explosive, discussions also involved Dr. Scott Atlas, who was constantly being attacked and undermined by Dr. Fauci and Dr. Birx. At a certain point, around mid-2020, it was decided that any task force member speaking engagements would have to be vetted across government, including my office (and the communications office), principally so that the content would be aligned with President Trump's messaging and that the task force members used the most updated, trustworthy evidence so that the country would be optimally informed and not confused. In short, this was to optimally coordinate the message of the task force with what Trump wanted to be able to say on the campaign trail, as well as what the science would allow them to say when leading members of the task force, including Dr. Fauci, went on weekend talk shows.

I'd been making the argument for several months that school closures and masking (especially in children) were not supported by the evidence, which was showing that these restrictive policies were actually ineffective and harmful. For example, we knew of the data we'd been receiving from Sweden, where large schools had not shut down or used masks, yet had similar or reduced risk of spread and deaths than nations or settings that did shut down and employed such measures. As an example, an email came to the highest government offices that included my office and me, which advised that Dr. Fauci was going on a talk show tour in the near future to discuss why schools had to remain closed and children had to remain masked. Again, the goal was to align the president's messaging at any given time with the task force's messaging so that this would engender confidence in the public. In response to the advisement of Dr. Fauci's planned talk show appearances, I informed everyone in the email chain that Dr. Fauci, the National Institutes of Health (NIH), and the Centers for Disease Control and Prevention (CDC) were basically wrong, because up to that moment, there was no science globally that supported school closures and masking in children or adults. I knew the email would attract great attention, as Dr. Fauci had been advocating for schools to remain closed and children to be masked, despite the science. I informed Dr. Fauci, the NIH, and all those present on the chain that they were wrong on the restrictive measures, and I shared the most updated science with them.

I ensured in the exchanges that Dr. Fauci and his team were included, and the follow-up exchanges (including in person) evidenced to me that leadership at the CDC and the NIH, as well as Fauci and his team, were very angry that I would openly school Dr. Fauci and the NIH. By this time, I'd had opportunities to interact with Fauci. One such instance was during "murder boarding," which is a technique devised to help prepare officials before they go to Congress for hearings. During the "murder boarding," Dr. Fauci would get angry when preparatory or anticipatory questions he did not like were posed to him by the mock congress and senate officials. When I asked Dr. Fauci questions, he would react with anger, even telling the attending leadership that he was not going to answer my questions. He did the same for others too.

Yet I never got that reaction from Dr. Francis Collins, then director of the NIH. I did not get that from Dr. Brett Giroir, Dr. Stephen Hahn, or Dr. Robert R. Redfield. Dr. Fauci was indeed different. I always found him full of hubris and arrogance given that we were only trying to help him prepare.

While I respected Dr. Fauci in principle, the best description I can give of him is that he was a diva, a star around which all the other planets, President Trump, and the country itself must orbit. Fauci himself has listed Mario's Puzo's mafia novel *The Godfather* as his favorite book, and, according to a *Business Insider* profile from 2021 about a glowing Disney/National Geographic documentary on Fauci, cites a single brutal line as his guiding philosophy:

> "It's not personal, Sonny. It's strictly business."
>
> The deadpan line, delivered by a young Al Pacino in the iconic 1972 film, *The Godfather*, has been a guiding principle for a different type of leader: Dr. Anthony Fauci . . .
>
> "When someone attacks, I don't immediately fight back. That's not my style. You don't get into the fray," Fauci says in the film. "And over the years, which became decades, that became the mantra, using *The Godfather* as the great book of philosophy: 'It's nothing personal, it's strictly business.'"[2]

As somebody who was an altar boy until I was eighteen years old and considered becoming a priest but for wanting to get married and have children, the idea that a man of science would worship at the altar of such an unholy philosophy is deeply disturbing.

And yet what I heard of Dr. Fauci and when I observed him, he seemed to act much like a Mafia don, surrounded by his "yes" men and women, never being caught with his fingerprints at the scene of a scientist's destruction or the failure of a promising treatment, and instead whispering orders to his faithful soldiers who would carry out his will.

In order to understand the near total control Dr. Fauci and his supposed superior at the time, Dr. Francis Collins, had over medical science, you have to put aside what you probably believe about the way science is done.

You probably think groups of scientists get together, discuss their research, and then come up with new approaches to investigate.

That's the way science is supposed to be done.

Don't we want our smartest minds posing challenging questions, maybe preferring the renegade, willing to strike out in a new direction, which might bring relief to millions suffering with disease?

But no.

Instead, we had Dr. Francis Collins and Dr. Anthony Fauci sitting on top of the medical funding pyramid. How big is that pile of money? I will provide you the answer from the NIH's own website in which they proudly declared:

> NIH received $41.6 billion in FY [Fiscal Year] 2020. Of this amount, $30.8 billion was awarded to 56 [&] 169 new and renewed meritorious extramural grants (excludes research and development contracts.) This investment was up $1.3 billion from FY 2019 (4.4 percent increase), with 1,157 more grants funded (2.1 percent increase) These awards were made to 2,650 academic universities, hospitals, and other organizations throughout the U.S. and internationally.[3]

Dr. Collins and Dr. Fauci essentially have more than thirty billion dollars a year to hand out to their friends and preferred collaborators. Looking across a thirty-to-forty-year period, one can speculate that this amounts to about one trillion dollars. How could one or two technocrats in the US government have that kind of allocation?

Do you have any idea how much influence thirty billion dollars a year can buy?

According to the website Open Secrets, political spending for the presidency in 2020 amounted to a little more than $5,700,000,000.[4]

Dr. Collins and Dr. Fauci control a yearly pile of money five times greater than that needed to elect a president of the United States.

And they have that money every single year.

There is no way in science you can escape the tentacles of control generated by this money.

Would ex cathedra Dr. Francis Collins and Dr. Anthony Fauci collude in order to prevent reasonable scientific debate?

According to the editorial board of the *Wall Street Journal*, that is exactly what they did during the COVID-19 crisis. A group of eminent researchers—Martin Kulldorff from Harvard, Sunetra Gupta from Oxford, and Jay Bhattachyra from Stanford—argued that the evidence favored the "focused protection" of high-risk populations such as the elderly or those with medical conditions. It was called the Great Barrington Declaration, and eventually was signed by thousands of researchers, including myself and Nobel Prize winner Mike Leavitt, also of Stanford University. As the *Wall Street Journal* editorial stated:

> In public, Anthony Fauci and Francis Collins urged Americans to "follow the science." In private, the two sainted public-health officials schemed to quash dissenting views from top scientists. That's the troubling but fair conclusion from emails obtained recently via the Freedom of Information Act by the American Institute for Economic Research.
>
> The tale unfolded in October 2020 after the launch of the Great Barrington Declaration, a statement by Harvard's Martin Kulldorff, Oxford's Sunetra Guta and Stanford's Jay Bhattacharya against blanket pandemic lockdowns. They favored a policy of what they called "focused protection" of high-risk populations such as the elderly or those with medical conditions. Thousands of scientists signed the declaration—if they were able to learn about it. We tried to give it some elevation on these pages.[5]

I was able to observe this duplicitous behavior while working behind the scenes at HHS for the COVID task force where the arguments I was making, supported with good data, were simply dismissed. I tried many times and in many ways to share information and discussions with all who would listen, even those officials who came to see me privately and anonymously, in fear for their safety and their careers. However, as you have seen and lived through during the COVID pandemic, in spite of clear evidence of the failures and harms of lockdown policies, the denial of early treatment when it was available, and the failure of the so-called COVID vaccines (gene therapy), officials like Dr. Fauci, Dr. Collins, Dr. Birx, CDC Director Dr. Rochelle Walensky, etc., are not pressed to answer for the failures. When questionable information becomes public knowledge, such as very serious questions that Dr. Fauci and Dr. Collins should have answered on the lockdowns, the COVID gene injection, and the origins of the COVID virus and how this disaster unfolded (regarding their direct roles in gain of function [GoF] research with the Wuhan Institute of Virology [WIV]), somehow they are able to elude proper investigation and accountability. I am routinely stunned that proper, public, and legal investigative actions were not taken against these two men (and others). Is it possible that, with the vast amount of money they have directed toward various researchers and institutions, few, if any, will rise to call for further investigation of these officials who lead the alphabet public health agencies and research and regulatory agencies, and those who report to them) and, if crimes are found, to prosecute them to the fullest extent of the law? The *Wall Street Journal* editorial concluded:

> Focused protection of nursing homes and other high-risk populations remains the policy road not taken during the pandemic. Perhaps this strategy wouldn't have prevailed if a debate had been allowed. But it isn't enough to repeat, as Dr. Collins did on Fox News Sunday, that advocates are "fringe epidemiologists who really did not have the credentials," and that "hundreds of thousands would have died if we followed that strategy."
>
> More than 800,000 Americans have died as much of the country followed the strategy of Drs. Collins and Fauci, and that's not counting the other costs in lost livelihoods, shuttered businesses, untreated illnesses, mental illness from isolation, and the incalculable anguish of seeing loved ones die alone without the chance for a family to say goodbye.
>
> Rather than try to manipulate public opinion, the job of health officials is to offer their best scientific advice. They shouldn't act like politicians or censors, and when they do, they squander the public's trust.[6]

All I can say is that the potential wrongdoing and apparent crimes of persons like Dr. Fauci and Dr. Collins and others in similar capacities are akin to generals of the Vietnam or Afghanistan wars telling the public for years that "there's light at the end of the tunnel in this conflict," "we're turning a corner," "as US forces stand down, the forces of our partners will stand up," or any of the other comforting bromides that are fed to the American people to justify the continuing loss of blood and treasure from our citizens.

* * *

I got to get up close and personal with Dr. Fauci in my role of helping him prepare for his appearances before congress, where people like US Senator

Rand Paul and US Senator Ron Johnson could be counted on to ask him challenging questions. These sessions, as mentioned, were called "murder boards" and were designed to anticipate all the questions which might come their way. They started in July and August 2020 after Dr. Fauci and other officials had stumbled badly in oversight hearings about his management of the pandemic response. I always felt that congressional oversight was critical. We would arrange the room to mimic a congressional or senate hearing room, with Dr. Fauci (or other officials) at the table with us (or on teleconference), and those of us questioning them from some distance away taking on the roles of congressmen or senators.

One time I remember asking him, "Dr. Fauci, there are reports you had a direct role in funding gain of function research at the Chinese lab in Wuhan, China. Can you categorically state that you had no role in providing taxpayer money for gain of function research at the Wuhan lab?" Gain of function research involves taking a pathogen and mutating (inserting genetic codes) into it so it has a new aspect to it, such as increased transmissibility or an increased ability to kill the organism it has infected.

Fauci would fly off the handle, proclaiming, "I'm Dr. Fauci! I don't have to answer your questions. And besides, I've answered that question before."

It shouldn't have been an unexpected question since *Newsweek* had reported on just such a possibility in late April 2020, just over a month into the crisis.

> In 2019, with the backing of NIAID, the National Institutes of Health committed $3.7 million over six years for research that included some gain-of-function work. The program followed another $3.7 million, 5-year project for collecting and studying bat coronaviruses, which ended in 2019, bringing the total to $7.4 million.
>
> Many scientists have criticized gain of function research, which involves manipulating viruses in the lab to explore their potential for infecting humans, because it creates a risk of starting a pandemic from accidental release.
>
> SARS-CoV-2, the virus now causing a global pandemic, is believed to have originated in bats. U.S. intelligence, after originally asserting that the coronavirus had occurred naturally, conceded last month that the pandemic may have originated in a leak from the Wuhan Lab.[7]

I simply have to point out that not many average people, or scientists for that matter, are convinced about scientific information when it comes from "US intelligence" regarding the coronavirus (from which COVID came) and its origins. Most of us would like more, shall we say, independent sources.

Even the former director of the CDC, Dr. Robert Redfield (who was an unexpected ally among those of us trying to pursue a more sensible policy, and I have to admit I thought a very decent man), came to believe SARS-CoV-2 escaped from a Chinese lab, as he told CNN in March 2021:

> Dr. Robert Redfield, the former director of the Centers for Disease Control and Prevention, told CNN he believes the coronavirus originally escaped from a lab in Wuhan, China. But it's too early to know for sure and investigations are ongoing.
>
> Redfield stressed he was not implying "intentionality," and no credible scientist, including Redfield, believes the virus was man-made.
>
> Still, Redfield's comments sparked debate. "I am of the point of view that I still think the most likely etiology of this pathogen in Wuhan was from a laboratory, you know, escaped," Redfield told CNN's Dr. Sanjay Gupta during an interview taped in January [2021], to be aired in full Sunday. "Now, other people don't believe that, that's fine. Science will eventually figure it out. It's not unusual for respiratory pathogens that are being worked on in the laboratory to infect the laboratory worker.[8]

In my interactions with Dr. Redfield, I found him to be a person of deep religious faith, who was interested in doing the right thing and, in my opinion, was not trying to subvert President Trump. However, he seemed to have difficulty asserting his will over his own agency.

In July 2022, Dr. Fauci's partner on the COVID-19 task force, Dr. Deborah Birx, would be telling a different story to the *Daily Mail* in the United Kingdom than she did from the podium as a member of Trump's team in Washington, DC. Not only did she also claim the virus likely escaped from the Wuhan lab, but also that its immediately virulent nature suggested it had been a lab creation:

> Infectious diseases expert and former presidential COVID adviser Dr. Deborah Birx told *The Mail* on Sunday that coronavirus 'came out of the box ready to infect' when it emerged in the Chinese city of Wuhan in December 2019.
>
> The adviser said most viruses take months or years to become highly infectious to humans. But, Dr. Birx said, Covid 'was already more infectious than flu when it first arrived.'

> She said that meant Covid was either an 'abnormal thing of nature' or that Chinese scientists were 'working on coronavirus vaccines' and became infected.
>
> 'It happens, labs aren't perfect, people aren't perfect, we make mistakes and there can be contamination,' she said.[9]

At the time of this writing, it's estimated that more than 6,400,000 people have died from COVID-19.[10]

Maybe it's just me, but Dr. Birx seems to equate the carelessness associated with causing the death of more than six million individuals to a bunch of teenage boys horsing around the house with a football and accidentally breaking a lamp.

I must add one other piece of information to this discussion, especially since Jeffrey Sachs, the head of *The Lancet*'s COVID-19 commission and director of the Center for Sustainable Development at Columbia University, has come out in support of this opinion.[11]

Senior officials at the WHO in Geneva and PAHO told me, in January 2020, about their suspicion of lab involvement in the creation of SARS-CoV-2, shared with me their opinion that this was done potentially under the umbrella of a biowarfare program in Wuhan, China, and said that they did not believe the release of the pathogen was accidental. They shared this with me confidentially, based on our work relationships and their understanding that I had completed a short program in 2001 at Johns Hopkins University in the area of bioterrorism and biological weapons research. As part of that program, I had the privilege of being lectured and taught by Dr. Donald Henderson, who was the architect of the eradication of small pox. I became friends with Dr. Henderson since I was considering his supervision in reading for a doctorate at Johns Hopkins in biological weapons research. Additionally, around July 2020, senior officials from NIH and HHS told me confidentially that they were also convinced SARS-CoV-2 was lab engineered, although they had no opinion as to whether the release was intentional or accidental. I was convinced, based on all I knew, that COVID was lab manufactured. The key issue has always been who the key players were and whether it was accidentally or deliberately released. I felt the coronavirus functioned almost identically to that of a bioweapon.

In a long interview with the magazine *Current Affairs*, Dr. Jeffrey Sachs gave his views on the origin of SARS-CoV-2, including the American contribution to the creation of the virus:

Now, what is the alternative hypothesis? The alternative hypothesis is quite straightforward. And that is there was a lot of research underway in the United States and China on taking SARS-like viruses, manipulating them in the laboratory, and creating potentially far more dangerous viruses. And the particular virus that causes COVID-19, called SARS-CoV-2, is notable because it has a piece of its genetic makeup that makes the virus more dangerous. And that piece of the genome is called the "furin cleavage site." No, what's interesting, and concerning if I may say so, is that the research that was underway very actively and being promoted, was to insert furin cleavage sites into SARS-like viruses to see what would happen. Oops!

Well, that is what *may* have happened. And what has been true from the start is the very real possibility, which a lot of scientists know, has not been looked at closely, even though it's absolutely clear that it could have happened that way. They're not looking. They just keep telling us, "Look at the market, look at the market, look at the market!" But they don't address the alternative. They don't even look at the data. They don't even ask questions. And the truth is from the beginning, they haven't asked the real questions.[12]

What Sachs was saying was similar to what my sources had been telling me right from the start of the pandemic, as well as what was being told to me months later by people inside Trump's COVID-19 task force.

And how bad was the stonewalling of information by the NIH when *The Lancet* commission was trying to get to the bottom of the issue?

The most interesting things that I got as chair of the *Lancet* commission came from Freedom of Information Act (FOIA) lawsuits and whistleblower leaks from inside the U.S. government. Isn't that terrible? NIH was actually asked at one point: give us your research program on SARS-like viruses. And you know what they did? They released the cover page and redacted 290 pages They gave us a cover page and 290 blank pages! That's NIH, for God's sake. That's not some corporation. That is the U.S. government charged with keeping us healthy.[13]

I wasn't having much better luck in my interactions with Dr. Fauci, although it was enjoyable to watch and listen to him get so angry.

When Dr. Fauci balked at answering my question (or similar questions) about funding the gain of function research at the Wuhan lab and threatened to not answer and essentially walk out, I patiently told him, "No, Dr. Fauci, you have not directly answered the question I have just asked."

But Dr. Fauci continued to resist. "I don't have to answer Dr. Alexander's question," he protested to others in the "murder boarding," looking for support from the other members present.

The moderator replied that I had asked a good question, and Dr. Fauci needed to answer it. He mumbled something non-responsive, and we continued.

There was always this effort to pacify Dr. Fauci since we were under constant threat from him directly and from his team and handlers that they would leak to the press that the White House, President Trump, HHS, others, and I were trying to muzzle him. No one ever did that, yet we were constantly threatened. This constrained and impacted dealings with Dr. Fauci as he, in his diva ways, ensured that he always got his way. As Dr. Fauci said in his documentary in which he asserted his love for mafia philosophy, I was sure that he would bide his time and figure out a way to kill my career and reputation.

It wasn't personal; it was just business.

But it wasn't the type of business which served the American people, or anybody looking for honest scientific debate.

* * *

On September 8, 2020, I exchanged a series of emails with Dr. Andrea M. Lerner, a medical officer at the Office of the Chief of Staff for Dr. Fauci's National Institute of Allergy and Infectious Diseases (NIAID),[14] which would set the stage for my removal (i.e., request for resignation or overt firing) from the US government.

It began with a request from Woleola Akinso for input about an upcoming appearance Dr. Fauci had scheduled for MSNBC. September 8 was a Tuesday, and we (all senior officials in government and particularly the alphabet agencies) were requested to respond with our thoughts by Friday, September 12. It read:

> Key messages/talking points: MSNBC would like to invite Dr. Fauci on for a live 10-minutes TV interview to go over some of the latest COVID-19 updates. They will have questions about vaccines, pandemic hotspots, & school reopenings. Dr. Fauci will provide them with updates on progress that has been made toward a COVID-19 vaccine. He will explain some of these factors which contribute to the emergence of "hotspots," and will discuss the factors that schools must take into consideration when deciding to reopen safely. He will emphasize some of the things that people can do on

an individual level to mitigate the spread of the pandemic, such as wearing masks, frequent and appropriate handwashing, maintaining appropriate social distancing, and doing activities outside when possible.[15]

The initial email was sent at 1:04 p.m., and I responded at 1:35 p.m. (via a series of back-and-forth emails between myself, senior NIH officials, and Dr. Fauci's team, as I understood it at the time) on the same day. One of my responding emails was the following:

Hi, my comment is this:

Can you ensure Dr. Fauci indicates masks are for the teachers in schools? Not for children. There is no data, none, zero, across the entire world, that shows children, especially young children, spread this virus to other children, or to adults or to their teachers. None. And if it did occur, the risk is essentially zero. It is the teachers who spread to teachers and take it home as per evidence. Children are more at risk from their teachers and teachers must engage in mitigation and mask and stay home if they are compromised and use a remote model, not the kids. The data is firm on this.

Moreover, the data is very firm on the fact that homes are higher risk settings than schools and are more likely to include older family members that must be protected from teachers, not children. The data also consistently shows that homes are where most cases spread, verified by data throughout the world. No doubt, the good news is COVID spares our kids and does not drive infection like seasonal influenza that is spread by children and kills way more children than COVID. The data is clear on that, too. Can Dr. Fauci contextualize this (COVID vs. influenza) as we now have 6 months of data so that parents are well informed by the actual data and evidence and not the media?

Dr. Fauci must be clear with this mask issue and provide the data that he is using to make this request to mask children. Did I read it right that Dr. Fauci is advocating masks for kids in school? It is just not so and this is flawed guidance if this is what he is calling for. Maybe I am missing something?? Can I see the data Dr. Fauci or anyone at NIH is looking at to advocate masking our children? Thanks.[16]

Imagine how different our world would be today if the data had simply been followed. Children would not have been required to spend that extra year at home, and the greatest inconvenience would have been to the teachers who remained masked for their own protection. The danger was *not* coming from the students.

If the actual data had been followed, students would have, for the most part, ended the 2019–2020 school year on March 14, 2020, and then been back in school for the 2020–2021 school year.

I have to spend some time talking about what it was like to be in support of the task force while at HHS and to have the opportunity to work with and interact with officials such as Dr. Hahn, Dr. Redfield, Dr. Giroir, etc. In my position at HHS, given close interactions with the communications arm of the government, it allowed me to be exposed to the machinations from HHS to the White House, as well as from the White House to HHS. I am unfortunately constrained due to sensitivities and confidentiality in terms of what I can openly share. I fully respect confidentiality, and I know you as the reader understand this. Yet I *can* talk about the reports we were getting about the devastation caused by the lockdowns and from children being kept out of school. I believe this is why President Trump was so incredibly passionate about trying to get children back into the classroom. I want the reader to understand that the lockdowns, business closures, school closures, shelter in place orders, and mask mandates all catastrophically failed. There is no country, no location, no setting, and no state globally where any lockdown or other restrictive policies worked to curb transmission or reduce death. They all failed.

First of all, we know that masking children all day is highly dangerous to their development. Young brains need the full amount of oxygen to properly develop. There was a reason chronic obstructive pulmonary disease (COPD) patients had a medical exemption. They need all of their oxygen as well. It's a basic fact of biology, in general, and human physiology, specifically. Also, many masks had tiny plastic particles that would detach and enter the students' lungs. Social development was also stunted because we like to see the faces of other people. And all of this was unnecessary because there was virtually no risk to children. My own comprehensive review (published by Brownstone Institute)[17] with other scientists on the effectiveness of the blue surgical and white cloth or masks, revealed that they are/were highly ineffective in stopping infection or transmission, as well as very harmful overall.

We were getting information about students at home who were committing suicide. Parents were finding their children—eight, nine, ten years old—hung in their bedrooms because of their anxiety, duress, depression, and unhappiness because of the lockdowns and school closures. The abuse of children was also skyrocketing. Day in and day out we'd get reports of parents who'd lost their jobs and were taking out their anger on each other and on their children, and then bringing those children into the emergency room, unresponsive in their arms, and saying, "I think I may have killed my child."

Physical or sexual abuse of children often comes to our attention for the first time in the school setting, and in closing schools, tens of thousands of cases were missed.[18] Children get their eyes and ears tested for the first time in school. When schools were closed, this could not be done. Millions of American children, and particularly poorer minority children, get their only meal in school (i.e., school lunch). By closing schools, millions of American children went for months with no lunch. We were getting reports daily from all states indicating the devastating effects of the lockdowns and school closures, where business owners, laid-off employees, and children were taking their own lives. We sent these to the White House, and I can tell you that President Trump was very angry, moved, emotional, and determined to reverse the ill effects. He fought with Dr. Fauci, CDC officials, NIH officials, and Dr. Birx, as well as with the teacher's unions. I can say the contention and battles were fierce, and he fought them hard as they conspired against him, subverting him at each turn in the pandemic response.

I was constantly trying to communicate information, data, and science to the CDC and the teacher's unions, but they didn't seem to be interested. And there were other considerations as well. As mentioned, children often get their ears and eyes tested for the first time in school, which identifies any problems. In addition, as mentioned, the vast majority of child abuse cases are first identified by teachers and counselors at schools. I was privileged to be in several teleconferences where we were informed that President Trump was present as this type of devastating lockdown damage information was being conveyed to him (and his advisors), and I could sense how much the information deeply affected him. It was clear that President Trump wanted those children back in school, where they were more likely to have their problems diagnosed and be safer, and thus the stress in the home would be greatly reduced. In my experience and all I knew, saw, and heard, I can state that President Trump genuinely tried to get America out of the lockdown lunacy that Dr. Fauci and Dr. Birx created, led, and implemented. Dr. Scott Atlas also fought Dr. Fauci and Dr. Birx to do the right thing and actually follow the science that he was giving them and informing them of near daily.

Dr. Scott Atlas arrived in August 2020 to the White House but quickly drew condemnation from the task force members and Deep State (I define the Deep State as the bureaucracy that stays as presidents come and go and considers themselves the true government). In my opinion, he was absolutely brilliant and just what was needed to bring the policy balance that was clearly missing. At that time, and now at the time of writing, October 2022, not one reputable, properly developed, cost-benefit/cost-effective analysis had been done on the COVID policies. This was Scott's expertise. He was a medical doctor with an extensive practical background in evidence-based medicine. The hatred and attacks on him by Dr. Fauci and Dr. Birx made me understand he was on target and doing the right thing. He was purely evidence and data driven, and they were out of their depth with him. They hated that and attacked him because he was so well prepared and informed. I can recall one of my meetings with Scott at the Eisenhower Building at the White House. I was so taken aback at how well prepared he was, as well as impressed with his grasp of all issues on early treatment and the lockdown response and his capacity to know data and numbers at the drop of a hat. *Impressive*, I would always think, and he struck me always as someone who deeply loved America and was doing all he could in his service to fix the stumbling and gross errors of the task force (in my view, the catastrophic damage the task force was doing). The nation needed a Scott Atlas. President Trump needed a Scott Atlas.

Dr. Atlas, like me, was hammering the CDC, NIH, and Dr. Fauci on school closures and the catastrophic effect they were having on children. As mentioned, we knew that one of the most important bulwarks protecting children from abuse were the schools. Teachers are the first line observers of abuse and mistreatment of children under their protection by their mandated reporting status when they observe mistreatment and abuse of children.[19] Scott argued vehemently to Dr. Fauci, Dr. Birx, and company that by closing schools, our public health officials intentionally severed a key avenue for detecting and intervening in the abuse of children who were homebound. We were shocked by the unscientific and harmful positions of Dr. Fauci and Dr. Birx. It was the first time on record in Western society where we reversed positions with children and asked them to suffer in order to protect adults from an infectious disease no worse than the annual influenza. We, in effect, made them into human shields.

On September 8, 2020, at 2:44 p.m., Jennifer Routh, the scientific communications editor at NIAID, responded to my thread:

> . . . Questions regarding the latest scientific evidence and resulting recommendations for schools are best directed to HHS, CDC and NIH scientific

staff. As a Task Force member, Dr. Fauci is aware of and communicates the recommended guidelines and consideration for schools, which have been formed based on available scientific evidence.

- Operating schools during COVID-19: CDC's Considerations
- Guidance for K-12 School Administrators on the Use of Masks in Schools

Jennifer Routh

News and Science Writing Branch

Office of Communications and Government Relations

National Institute of Allergy and Infectious Diseases (NIAID)[20]

I quickly wrote back and sent the following email at 3:31 p.m. to the task force (as an example of the back and forth email communications):

Hi Jennifer, very respectfully and thank you for the timely reply.

The studies referred to in the guidance you shared are not applicable to my statements or questions. They pertained to older folk, family who were older, long term residents, and Diamond Princess.

Dr. Fauci has no evidence, zero, that children spread this virus to children in schools or to adults. This is a global situation. And no evidence that masks are needed on young children. None. Zero. It is wrong for it is not only traumatic to them, there is no evidence. So why do that when there is no evidence? I never said not to engage in reasonable mitigation steps but mask the teachers etc. but not the low risk children, at essentially zero risk of severe illness or death. There is strong stable evidence that children are the seat and driver of influenza but there are no requests to mask kids annually for that or did I miss that? There was no request to mask children by the CDC when H1N1 in 2009 ravaged children or did I miss that too? I may have. I shared a study specific to that showing infection from COVID comes from the home, not even school. That too is clear.

I ask again, where is the evidence the NIH or CDC or anyone uses to support masking children? There is none. That's what I ask about. So I am asking for it and simply ask Dr. Fauci to contextualize COVID relative to seasonal influenza so that parents and the public are informed not by the media and unsubstantiated information. Tell the public how many children die of COVID vs. influenza. Tell the public how many get severely ill and this will be helpful. Not how many infections there are, for while interesting, it is not what matters to parents. Thousands of kids get influenza infection daily.

Please provide data to me and I will be corrected. Dr. Fauci is a lead spokesperson (my deep and humble regards to his expertise) and if he makes a statement, he knows it must have data behind it. None of the actions advocated for on our children have even been done for influenza yet influenza is more devastating to our children and each year. Why have we not masked them? They take influenza home but do not take COVID home. I operate on data and applicable data.

I checked each of the studies you referred to and they are not applicable to this discussion.

Thank you.[21]

I was doing my best to keep a civil tone (I counted four times in that email I expressed my appreciation for their hard work and experience and that I was happy to change my opinion when presented with new data), but we were changing the lives of hundreds of millions of Americans without good data. The high-quality, optimal, trustworthy evidence was never there to justify the restrictive lockdown COVID polices being enacted, and the more people like me raised this, the more we were attacked, smeared, and threatened with leaks and termination. In fact, we knew clearly that these lockdowns were killing children, with additional suicides, were ramping up domestic abuse, were inflicting long-term psychological damage, and were causing kids to fall behind in school. Yet it became apparent to me that those in the administration, on both sides, and particularly Deep State-entrenched bureaucrats and technocrats, were not interested in the "real" science.

At 5:07 p.m. on the same day, I received a longer email from Dr. Andrea Lerner, which also went out to all the other members of the task force. It read:

Dear Dr. Alexander,

I am an infectious diseases physician on Dr. Fauci's staff. While transmission dynamics of SARS-CoV-2 involving children are not fully understood, potentially complex and probably differ among age groups, I don't feel it is correct to say there is "no evidence, zero, that children spread this virus to children in schools or to adults." Or that, "They take influenza home but do not take COVID home."

In addition [to] other benefits of reducing SARS-CoV-2 transmission in a community, it seems wise to use tools to prevent outbreaks in schools so that schools that do open have a greater chance of remaining open and kids can reap the benefits. As was pointed out, this is in line with CDC guidance.

Here are some relevant articles I have come across:

https:///www.cc.gov/mmwr/volumes/69/wr/mm6931e1.htm

A description of an outbreak at a Georgia overnight camp.

"The overall attack rate was 44% (260 of 597), 51% among those aged 6-10 years, 44% among those aged 11-17 years, and 33% among those aged 18-21 years."

https://wwwnc.cdc.gov/eid/article/26/10/20-1315_article

"In households with an index patient 10-19 years of age, 18.6% (95% CI 14.0%-24%) of contacts had COVID-19."

Where index case was 0-9 years of age, 5.3% of household contacts had COVID-19.

This illustrates that preventing outbreaks of COVID-19 in children in schools is important as they can transmit to their families when they come home.

https://protect2.fireeye.com/url?=5c00ee4e-0054f732 -5c00df71-0cc47adc5fa2-802e82c136d4663&u=https://eurosurveillance.org/ content/10.2807/1560-7917.ES.2020.25.29.2001352#html_fulltext

A description of an outbreak at an Israeli school (students aged 12-18).

"Testing of the complete school community revealed 153 students (attack rate: 13.2%) and 25 staff members (attack rate: 16.6%) who were COVID-19 positive. Overall, some 260 persons were infected (students, staff members, relatives and friends)."

https://www.sciencedirect.com/science/article/pii/S0022347603110234 ?via%3Dihub

"Nasopharyngeal viral load was highest in children in the first 2 days of symptoms, significantly higher than hospitalized adults with severe disease."

The authors hypothesize, "children with high viral loads and non-specific symptoms including rhinorrhea and cough can likely transmit SARS-CoV-2 as easily as other viral infections spread by respiratory particles. If schools were to reopen fully without necessary precautions, it is likely that children will play a larger role in this pandemic."

Very respectfully,
Andrea Lerner

Andrea Lerner, MD
Medical Officer
Immediate Officer of the Director
National Institute of Allergy and Infectious Diseases (NIAID)
National Institutes of Health.[22]

I genuinely appreciated this email from Dr. Lerner as I was finally getting some data upon which they were basing their decisions. However, my entire training in evidence-based medicine was predicated on looking at published data and determining whether they'd been making the correct comparisons and coming to a well-reasoned conclusion.

I finally felt as if somebody was engaging me in a scientific debate, and that this was a chance for me to display the skills for which I'd been recruited by HHS to support the COVID-19 task force in the first place.

> Dear Dr. Lerner, so very respectfully.
>
> What you shared is important to consider in the larger body of evidence and it is important we consider both sides. I also include evidence as attached above that informs me. [I included eight published articles.] Just a few but there are lots if you'd like additional evidence. Data is accumulated now to show that SARS-CoV-2 is spread mainly between adults and from adults to children. Not the other way around.
>
> I would also think you would find the Danis Alps paper very instructive to this debate for it shows that an infected child did not spread to any of their close contacts in schools (over 150 or so). Importantly, no additional cases were identified within the following 14-day follow-up period of all the contacts. The public needs sound honest answers and the data and evidence does not support masking children. It is important for parents to know this. It's a simple argument I make for it is traumatic to the child and if it confers no benefit, why do it? Not on what we feel or think, but what the data shows. Moreover, in an attached extensive review looking at spread among children in schools (all studies), researchers found:
> - Children are not a major source of transmission of COVID-19.
> - Analyses of infection clusters revealed that for children who were infected, transmission was traced back to community and home settings, rather than among children within daycares or schools; children did not spread it among themselves.
> - Within household clusters, adults were much more likely to be the index case than children.
> - Prevalence of COVID-19 infection in children in daycare and school settings was lower than the prevalence of COVID-19 in adults working in daycare and school settings.
>
> Respectfully, we can go back and forth and I too agree with you that the key is to reduce risk of transmission to the vulnerable high-risk group. We have to protect our precious elderly. I however stand by my statements below

fully based on attached evidence and more I am conversant with. There is little if any evidence (and in our statistical world we even say 'o' when it is that negligible) that children spread this virus and take it home and what exists is very limited and I can take it apart methods wise.

I guess because we have had seasonal influenza for decades and no effort to mask children when we know children drive it from schools, and because we did not mask for H1N1 2009, I question why the push to mask kids here for COVID when the overwhelming evidence suggests that spread comes from home. We even have evidence nascent as it is, that children may not have the ACE2 receptors needed in the first place to get infected to then spread (see Bunyavanich 2020 above). Children drive influenza and we have never masked them, and they take it home and cause serious illness to elderly, but we do not mask them. Why now? This is an important question that no one seems willing to answer.

Let me end by sharing this:

The great Dr. John Ioannidis [Stanford University], whom I confess I learn from daily, has summed up what the entire world's data and science demonstrate conclusively: the risk for children and young adults having severe illness and dying from COVID-19 is "almost zero." This simple statement is what parents need when we talk about their very low risk children. And if COVID for kids is framed relative to seasonal influenza which harms kids, by someone like a Dr. Fauci, then parents will understand it for the first time.

We can disagree respectfully, and I do with Dr. Fauci or whomever on this matter if he advocates for children masking, while in very deep respect of his career. On balance, there is no evidence to mask children. If the argument is they could take it home, the risk is not there from kids. That extent of risk. I do not mean zero risks like 'o' but very negligible risk and I will use Dr. Ioannidis terms 'almost zero' and I think you understand that. I see other strategies being quite useful but my view is masks can have negative effects on the kids akin to how I argue the lockdowns and sheltering and masks can/ would have a damaging impact on the immune system of children/adults if prolonged. Something to think about.

Thank you for sharing and thank you for the very fine work you do.

Very respectfully,

Dr. Paul Alexander, PhD

Senior Advisor to the Assistant Secretary for COVID-19 Pandemic policy

Office of the Assistant Secretary of Public Affairs (ASPA)

US Department of Health and Human Services[23]

I felt I'd done an excellent job of explaining my position to the COVID-19 task force, especially including the eight articles I'd included with my email, calling special attention to two of the articles in particular: the Danis paper regarding an outbreak in the French Alps published in the *Journal of Clinical Infectious Diseases* and the Supinda Bunyavanich paper from the *Journal of the American Medical Association* on why children accounted for less than 2 percent of identified COVID-19 cases. There were many other electronic exchanges along the same lines where I argued with the science against the lockdowns and school closures as well as the ineffectiveness and harms of face masks.

Specifically, in the Danis paper, a young child (but not his two siblings) was infected with the virus at a chalet in the French Alps, and then attended different schools where he interacted with more than 150 people, none of whom became infected.

Particularly, the infected child, despite interactions with a large number of contacts in different schools, did not transmit the disease, as evidenced by the large number of negative results of his tested contacts. However, the high proportion of picornavirus and influenza infections among his contacts at the schools indicated transmission of those viruses within those settings. Similarly, we observed that the family cluster allowed the dissemination of picornaviruses on influenza A viruses in the 3 children, while SARS-CoV-2 was detected in only 1 child.

These 2 observations suggest that picornavirus and influenza infections are more easily transmitted than SARS-CoV-2. It is possible that viral interference in the host may impact the individual susceptibility to another viral respiratory infection as observed during the 2009 influenza pandemic and other winter seasons between A(H1N1) influenza virus and respiratory syncytial virus. It is also possible that the very low viral load of the pediatric case and the subsequent lack of transmission might be related to his coinfection and the co-circulation of respiratory viruses . . .

. . . The child continued his normal activities and interactions as his symptoms were mild. Current evidence indicates that children develop COVID-19 less often than adults and the clinical manifestations of the disease are milder. The above suggests that children, being less likely to be infected and more likely to develop mild disease, may play a less important role in the transmission of this novel virus.[24]

This paper had been published on August 1, 2020, meaning they'd had several months since the February outbreak to review the data and check their findings. In short, the child had only mild symptoms, did not infect his two siblings, and also did not infect any of the more than 150 students and staff with whom he was in close contact at different schools. This French Alps data supported what had already accumulated since February and March 2020, which was that children did not readily become infected, did not transmit it to other children, and were not drivers of the infection homeward; that children were actually more at risk from their teachers.

The Bunyavanich paper was especially interesting because it sought to answer one of the more puzzling questions of SARS-CoV-2 infection and COVID-19—namely, why children made up only 2 percent of the cases and seemed to have very mild symptoms. And it was a robust study, with 305 individuals, aged four to sixty. One must understand the point of entry for most respiratory viruses is the nose, and it was discovered that point of entry for the virus into the cells was the ACE-2 receptor (angiotensin-converting enzyme),[25] which was not as active in children as in older persons. From the paper:

> Children account for less than 2% of identified cases of coronavirus disease 2019 (COVID-19). It is hypothesized that the lower risk among children is due to differential expression of angiotensin-converting enzyme (*ACE2*), the receptor that severe acute respiratory syndrome coronavirus 2 (SARS-CoV-2) uses for host entry. We investigated ACE2 gene expression in the nasal epithelium of children and adults.[26]

In my estimation, this was a ground-breaking paper that explained the enormous difference we were observing in infection rates between children and adults. I was also well aware of the strong research and body of evidence (and assumed that the CDC and NIH, as well as Dr. Fauci, Dr. Birx et al., and their teams were also aware) showing that children's innate immune systems (the first line of immunological defense) were pre-activated (primed) in the upper airways and worked to control early SARS-CoV-2 infection. This natural innate immune system and response are more primitive, broad, non-specific, potent types of responses pre-activated and primed early after infection. The evidence was clear that the innate immune system especially was comprised of immune molecules that were geared to mobilize and combat viral replication.

It wasn't that children weren't being exposed to SARS-CoV-2 viral particles. It's just that they couldn't get much of a foothold in children because their cells blocked entry of the virus. It was only within the cell that a virus could make more copies of itself. This partly explained why children were, and are, less likely to be infected in the first place, spread it to other children or adults, or even get severely ill since the biological molecular apparatus is simply not there in the nasopharynx of children.

Overall, the existing and rapidly accumulating body of evidence was showing us clearly that children were protected against SARS-CoV-2 virus, and that COVID essentially spared our children. We knew soon after lockdowns began in March 2020 that children generate a robust, cross-reactive, and sustained immune response to SARS-CoV-2 with focused specificity for the spike protein. Somehow Dr. Fauci, Dr. Birx, their teams, and supporting scientists at CDC, NIH, NIAID, and even the Food and Drug Administration (FDA) were not reading the science, "getting" the science, or understanding the science. I grew to think that there was tremendous cognitive dissonance within these agencies to any narrative disparate to theirs and that these alphabet health agencies operated within an atmosphere of "extensive academic sloppiness."

The discussion section of the Bunyavanich paper really put it all into proper perspective:

> The results from this study show age-dependent expression of *ACE2* in nasal epithelium, the first point of contact for SARS-CoV-2 and the human body. Covariate-adjusted models showed that the positive association between *ACE2* gene expression was independent of sex and asthma. Lower ACE2 expression in children relative to adults may help explain why COVID-19 is less prevalent in children.[27]

I thought I was doing exactly what I was supposed to be doing in my role as an evidence-based expert for the COVID-19 task force.

However, at a meeting shortly after that email exchange I was going to get a very rude surprise.

* * *

My office at HHS was on the sixth floor. Just down the hall from me was the office of Alex Azar, Trump's pick to head the HHS agency, a former pharmaceutical executive who always struck me as technically adept yet very slick. I personally did not think he was the optimal person to head HHS during the pandemic response.

A set of nearby stairs would take me to the seventh floor, headquartered home of "Operation Warp Speed (OWS)," Trump's operation to develop a COVID-19 vaccine in record time. I'd often talk with many of those OWS scientists, and they'd tell me of their concerns, saying the vaccine was being developed without the necessary safety precautions or timeframe to check for side-effects. It surprised me that they would confide in me and seek me out, yet I was told that because I was outspoken on the failures of the response, people, even those within the Deep State, gravitated toward me. They told me they felt they could "trust" me. They were always asking for my ideas and views on research methods and science, which took me aback. The reality is that I'd already developed a reputation as one of the voices challenging the proclamations of the COVID-19 task force, so I guess they sought me out as a fellow renegade. I also noted that the OWS office seemed to have an inordinate number of military officials working there. I grew to understand that OWS was as much a military operation as it was a vaccine development operation. Very intriguing to me was that, along with OWS being at home on the seventh floor just above my office, Moderna's vaccine development headquarters were also on the seventh floor just above my office. I also had several opportunities to interact and meet with Moncef Mohamed Slaoui, who was tasked by President Trump to lead OWS. Dr. Slaoui was one of the officials generally open to these discussions, but I grew to wonder if his loyalty was to Moderna or to science. I also interacted often with Dr. Hahn who headed the FDA. Dr. Redfield, who headed the CDC, was somebody I interacted with daily as he often shared his frustrations with me about agency politics. Dr. Giroir and I shared a lot of science, often reviewing breaking published papers. The intense collaboration between scientists is common in science, and at the beginning it was easy to believe we were trying to implement best practices, but as time went on, it seemed another agenda was at play.

I recall on the fifth floor or thereabouts of the HHS building was our dedicated conference boardroom, where we'd have meetings with other scientists, lawyers from the White House, and people from other agencies.

One of the surprising things about the meetings held in that conference room (and similar designated rooms) was the stark difference between how we were expected to act in the hallways and in the meeting rooms. In the hallways we wore our masks and kept a six-foot social distance. Inside the meeting room, the masks were removed, and we'd sit elbow to elbow. It was very fascinating to me.

I believe it was a day or two after the September 8, 2020 series of emails, but before Dr. Fauci made his appearance on MSNBC, that we had another meeting, and I noted some new faces. Such meetings were often overpacked with officials from all agencies in the government, people from the White House, Department of Justice lawyers, etc. Thus, to say it was "chaotic" is an understatement. These were very serious discussions at times, yet you had the sense that you were laying tracks down as the train was barreling forward full steam. After the meeting, they asked to speak with me privately as we walked to the elevator and me to my office. I decided to walk with them as I was always interested in meeting new people in government and learning what they did. This was the first time I saw them in meetings in person, but many meetings were held by teleconference.

In our walk, they told me they'd often heard me on the conference calls, as my distinctive Caribbean accent and bold manner of speaking made me stand out. It was sort of upsetting in that they alluded that people on the other side of the calls would always whisper and ask who the "Jamaican" guy who is the scientist at HHS is, etc. Some of what they shared around that was near discriminatory in content and is not in the scope of this book. They told me they often agreed with what I was saying but then got specific by telling me that I'd made a serious mistake by criticizing Dr. Fauci and the NIH openly in my email about school closures. In fact, they told me that CDC and NIH, as well as Dr. Fauci's camp, were upset about my direct communications to them and the overt manner in which I would tell them about their mistakes and misguided approach to the response, especially when I told them that they were not "science" based.

So, of course I was startled by what they were sharing and asked "Why? What did I do?" I then said, "I didn't mean to offend anybody. I was just speaking of science, and they shouldn't do things if they weren't supported by the data."

"Nobody questions Fauci," they said. "You never question him and by doing it openly, schooling him openly, then you made yourself a target for his shop."

I couldn't believe what I was hearing.

"You really don't understand what you've done, do you?" they asked me.

"No. There must be something I can do," I replied. "I can write a new email right now and apologize to Dr. Fauci for my tone, if he found it offensive. I am happy to send it to every member of the task force, for while

I stand by my science which is informed by the most updated data and evidence, my aim was not to insult. I am confused as to why he or they would feel this way."

"It is a bit complicated; by your schooling them with your thick Caribbean accent, your known 'volunteer' status [for a long while at HHS, the Deep State was refusing to complete my hire paperwork or actually pay me a salary so I was designated as a volunteer until this was addressed], your immigrant origins, talking openly at such a high level, the fact that they were not accustomed to a high-level academic scientist coming out of the HHS and communications, and the lack of your Ivy league schooling—that is part of it," they told me.

They also stated that high-level officials at the CDC, NIH, and FDA, who were often on the calls I attended, were recently questioning how I arrived at HHS and weighed in on the e-mail exchanges I had with NIH, with some saying "why would this volunteer person, this Black Jamaican island person who is a volunteer, this Black guy, a so-called scientist, think he can lecture Dr. Fauci and NIH, coming from a low village type unknown university in Hamilton, Canada, this non-pedigree university. We are not going to be lectured by a Black low-level-type scientist volunteer, and who listens to this volunteer anyway?"

I then replied in a joking sort of manner, "Is that how they regard me? At the CDC and NIH, such high-level officials? I regard them highly and have the greatest respect for each of them, so why would they behave in this manner to me? They do not even know me, and have not even met me in person. I am stunned and outraged for those are incredibly discriminatory and racist statements. They are also misinformed because those who could not gain entry into the research methods and evidence-based methods program that I completed my doctorate in at McMasters University, Hamilton, Ontario, then proceeded to Harvard and Stanford, etc. It is the best in the world and purists and advanced scientists know this. It is likely the technocrats were not familiar. My doctoral supervisor actually founded the entire field of evidence-based medicine. Anyway, how can I address this?"

They then informed me that, "they at NIH and CDC and Fauci will move to destroy your name and career now, and you will never be able to work in DC or recover. They will use the media to slander you and cancel you and censor you."

I then asked, "Please, tell me what I can do?"

They then continued, "There is nothing you can do. It's already been decided. They're going to cut your balls off. In a few days they're going to leak a few lines of an email of yours and build a story around it. The media is waiting and will mischaracterize it and write falsehoods so that your name will be dirt. You'll be a rogue, nut scientist working for Trump who wants children to die. These people are experts at destroying the careers of those they don't like. There's nothing you can do to stop this. We just wanted you to know what's coming because we really like you and the science-based informed things you've been saying. The truth is that it is not about you, Paul; this is about Trump. You work for him and his HHS so you are marked for destruction as are all who work there for him, so don't take what they are about to do to your name and career personally. Between NIH, CDC, Fauci, and the Deep State bureaucracy, they will now move to have you fired for challenging Fauci and NIH. Fauci and NIH will ensure you are fired."

They parted from me in the hallway and wished me good luck.

I felt like a man alone in a small boat in the middle of an ocean, knowing a category five hurricane was bearing down on him, and he'd just heard the first roar of the storm.

How could I possibly survive?

CHAPTER ONE

A Child of the Storm

My family has survived worse storms than the wrath of Dr. Anthony Fauci.

My father's side of the family is from Egypt, Syria, and Lebanon. I'm told that my great grandparents (father's side) sailed for the New World about a hundred years ago, but in the Caribbean, they ran into a hurricane which sunk their ship. Even though my great grandmother was several months pregnant, she and my great grandfather made it into a lifeboat and eventually landed on the shores of Cuba. My grandmother was their first child and was born in Cuba. You might say the legacy of that storm, and the determination of my great grandparents to survive it, lives on inside of me.

My mother's heritage, by all accounts, was part Venezuelan, part Portuguese, and a mixture of a multitude of other backgrounds. I learned from my mother and grandmother that my grandfather left for Venezuela when she was a child. Up to this day I remain unsure. My maternal grandmother told me that her great-great-grandfather was a plantation slave. This was a surprise to me; as she aged, we became very close, and I would go visit her and sit and talk for hours. While I thought at times she was showing varying grades of dementia as age set in, some of the stories were riveting, and I respected them as gospel. The knowledge of my heritage made me proud of my family's strength and ability to endure great hardships. When I consider that my ancestors survived terrifying shipwrecks and the degradation of slavery, it makes much of what I am about to relate in this story much easier to bear.

I was born on May 18, 1965, in a small town in Trinidad in the Caribbean. Technically, I was born in the Republic of Trinidad and Tobago, two of the southernmost of the Caribbean Islands, one section of Trinidad being located just seven miles from Venezuela. In one sense, I was fortunate because the country is the most prosperous in the Caribbean, yet my immediate family was not of means and struggled most of my life. It is one of the main reasons why I decided to go abroad to continue educating myself while helping my family financially. The system of education (primary and secondary) in the Caribbean, and in Trinidad and Tobago specifically, was based on the British Cambridge System and was at the top globally. This was because Trinidad and Tobago was a slave colony of many European nations and fell for a long time under British rule. The country was politically stable, and the strong Catholic tradition gave me an unshakeable underpinning of faith. I was raised with a strong Christian faith, yet I had many friends I cherished from other faiths, such as Muslim, Hindu, etc. Trinidad and Tobago has a strong Hindu faith influence due to the historic ethnic makeup.

I became an altar boy at a young age, serving in that role until I was nearly eighteen years old. Being an altar boy in Trinidad is what the promising young men were encouraged to do, and I can never forget the feeling of connection to the community I experienced. We were in charge of preparing the community church for weddings and funerals, making sure the flowers were properly arranged, laying the carpets, making the day memorable for those couples who were beginning a new life together, or saying goodbye to those whom God had called home. The altar boys would also often accompany the priests around the island as they visited people who were too sick to come to church or brought food to those who had fallen on hard times. I vividly recall the people I met when we accompanied the priest thanking us for coming, and realized the immense importance of connections between human beings. We live on this small planet together, helping when we can, and if there is nothing to be done, bearing witness to what we have seen, and asking God to show mercy toward those who suffer.

I can't recall the exact date, sometime between 1984 and 1986, we in Trinidad got cable television for the first time, and the world beyond my small island nation opened up to me. I was around eighteen or nineteen years old, the time in life when a young man ponders what type of man he wants to be when he grows up and what legacy he wants to leave for his children. I became obsessed with CNN, and particularly the American president, Ronald Reagan, who genuinely loved his country and was engaged in a titanic struggle with the atheist communists who ruled the Soviet Union and sought to purge the very idea of God from the world. I loved Reagan's deep belief in law and order and how he defended the flag of the United States and supported the military. I was mesmerized by Reagan.

Around this time, I also saw the movie *Top Gun*, starring Tom Cruise as Pete "Maverick" Mitchell, and this only deepened my love of the American military. I wanted to be one of those guys, the "best of the best" and put my life on the line for the greatest country in the world.

In retrospect it may sound foolish, but I sat down one night and wrote a letter to the secretary of the United States Air Force telling him that even though I lived in Trinidad, I wanted to serve in the US military and become a fighter pilot. Surprisingly, a few months later I received a letter back from the Air Force, informing me that the United States often allows foreign citizens to enlist in times of war, and in return those soldiers are eventually granted citizenship. But, as the United States was not engaged in any fighting at the time of my letter, that possibility was not open to me. I was disappointed to find myself unable to serve the United States, but the letter only served to increase my interest in someday living in America. I loved the Pledge of Allegiance and the idea of freedom and imagined myself someday defending the American flag. I even framed the rejection letter and hung it on the wall. I recall inviting all my friends to come see this rejection letter in a frame with the US emblem on it. I was very proud of it. Just to have the US government's emblem on a letter addressed to me.

When I was a young man in Trinidad, all my friends dreamed of going to the United States. I took to heart the admonition of Dr. Eric Williams, the first prime minister of an independent Trinidad and Tobago, who said "The future of our nation is in our children's school bags."[28] He told us to wear our island nation as the boys and girls we are and the young people we would grow to be. Eventually an opportunity came to study abroad, not in the United States, but in its neighbor to the north, Canada. I was twenty-three or so when I moved to Canada to begin my education. Maybe it was the fact that I was several years older than most of my classmates and that they didn't have my experience of living a hard life on a Caribbean Island, but when I received my first degree, I received a special award, the Governor-General's Medal of Canada for having the highest grade point average of all graduating university students that year. I was surprised that I did so well and credit my secondary education under the Cambridge System in the Caribbean.

One university course in particular (Medical Geography) put me on a different path. This course looked at the cardinal impact of geography and disease and how it has shaped the world's history as we know it. It focused on the political, demographical, ecological, and psychological effect of disease on global cultures (from smallpox in Mexico, to bubonic plague in China, to the typhoid epidemic in Europe, to HIV/AIDS globally) and the history of disease and the impact on humankind.

I'd always loved reading about history and geography, trying to understand why countries and leaders rose or fell. This course specifically looked at human history through the lens of health and disease, detailing what ailment a military or political leader might have had that changed the course of a battle or a political campaign. Then we also learned about the different health statuses of clashing civilizations, such as why Europeans with their guns and germs were able to overpower the great Mayan and Incan civilizations, and how, when the Americans traveled west, they were entering a devastated landscape where the native population had been drastically reduced by disease. The professor taught us about the symbiotic dance (always a quest to establish equilibrium) between viruses and humans, not a war to the death, but a process of assimilations where the pathogens sought to live in relative peace with the host, hopefully conferring some survival benefit, because that was the environment in which the pathogen hoped to continue existing. When the system became unbalanced, the pathogen could begin raging through civilizations and causing great damage.

One of the things we've come to understand is that the damage pathogens cause may not be a direct result of the virus or bacteria itself, but the body's overreaction to the invader. One of the reasons I write so passionately and dedicate my life to this crusade against Dr. Fauci and the corrupt public health system of the United States (spearheaded by the four Horsemen of the Apocalypse, alphabet health agencies CDC, NIH, FDA, and NIAID) is I believe we have witnessed exactly that scenario in the body politic. Dr. Fauci, Dr. Collins, and, in my opinion, their henchmen and women have provoked this violent societal overreaction and, in the process, caused massive unnecessary suffering. Viruses and humans both share this planet. We can live in peace with the pathogens of our world, but it requires us to understand that health is not simply the absence of disease. It is the presence of a robust immune system, exquisitely fine-tuned to repel biological and chemical trespassers.

Given my interest in contributing at some level to creating a more peaceful world, it may seem surprising that the next direction in my life was to take a course at Johns Hopkins University in the epidemiology of bioterrorism and biological warfare. Like many who begin to delve more deeply into this area, I became concerned with the possibility that some rogue nation could purify and weaponize a deadly pathogen, load it onto a missile, and fire it at an innocent population center. The department at Johns Hopkins was headed by Dr. Donald Henderson, who was world-renowned for the effort he'd led beginning in the late 1970s to eradicate smallpox from the globe.

After the first class, I nearly sprinted to the front to speak with him about his lecture. It is my nature that when there's something I want, I head toward it like a bloodhound. Our first Trinidadian prime minister also said to us that wherever we go, we must "wear our island on our sleeve." I always took that to mean we should walk confidently in the wide world and know that with enough hard work we could accomplish anything. Because I was constantly coming up to him after lectures, we soon became friends, and I emailed with him back and forth. Dr. Henderson eventually agreed to supervise my planned doctorate in biological warfare when I asked him to. When Henderson died in 2016, this is how he was eulogized in the *New York Times*:

> Dr. Donald A. Henderson, a leader of one of mankind's greatest public health triumphs, the eradication of smallpox, died on Friday in Towson, Md. He was 87.
>
> Dr. Henderson, who lived in Baltimore, died in a hospice of complications of a hip fracture, including infection with antibiotic-resistant staphylococcus, a dangerous pathogen he had himself researched and raised alarms about, said his daughter, Leigh Henderson.
>
> Starting in 1966, Dr. Henderson, known as D.A., led the World Health Organization's war on the smallpox virus. He achieved success astonishingly quickly. The last known case was found in a hospital cook in Somalia in 1977.
>
> Long after the disease was officially declared eradicated in 1980, he remained in the field as a dean of what is now the Johns Hopkins Bloomberg School of Public Health and an adviser on bioterrorism to several presidents.[29]

However, at the same time I was considering working on bioterrorism, I was also thinking about how to use my skills and found myself attracted to Oxford University and their masters in evidence-based medicine and clinical epidemiology. In the years since, I've often thought of that critical juncture in my life and how what seemed such a relatively inconsequential decision would so drastically alter the man I would later become. What would have happened to me if I'd pursued a doctorate in biological warfare and gotten the chance to work under someone like Dr. Henderson? My discussions with him already showed me his deep intellect and passion for doing good for America and the world. He was an incredible person if you had the chance to meet him. Truth is, the most likely outcome is I would have likely disappeared into the shadowy world of clandestine scientific work, maybe becoming one of the mad monks of the Defense Advanced Research Projects Agency (DARPA), convincing myself I was some secret soldier saving humanity from unimaginable nightmares, and yet wondering as I lay awake at night, pondering the lies I told myself and the things I claimed not to see, if I had become the very monster I sought to destroy.

At Oxford, I fell in love with evidence-based medicine. I saw it as "pure" science, looking at the evidence and research methods and critically appraising and evaluating what others had done, a final check on the work before placing it before the public. It was vitally important to me for the public to believe the science that was being placed before them. It must be free from politics or corporate interference (conflicts of interest) which seek to profit from the ills of humanity. Maybe it was those years of being an altar boy, and how religion teaches one to have an intimate relationship with God, where nothing is hidden. Evidence-based medicine seemed to me like the scientific equivalent of a private, honest prayer between me and my Creator. It didn't mean I couldn't be wrong. It simply meant I was doing the best that was humanly possible to do. And if somebody disagreed with me, I expected them to make a stronger argument, based on either better data or a superior analysis, to the one I had made.

While at Oxford, I reached out to my colleagues at McMaster University because I wanted to continue my evidence-based studies and obtain a doctorate. I became close with Dr. Gordon Guyatt and grew to know Dr. Dave Sackett, who are considered the fathers of evidence-based medicine. Gordon became my doctoral supervisor as well as my post-doc supervisor. Again, I was falling back on my island philosophy. I don't care how accomplished a person is: if they know something I want to know, I will go up and talk to them. I have no hesitation. I want to learn things and be helpful to the world. When we gained our independence from the United Kingdom in 1962, our prime minister said that the era of the slave master was over. We had once been colonized, but that time was forever gone. I was always told by my parents that the world was mine to conquer, but that I should always have the values of our island in my heart.

If it sounds like my life was all about work, it wasn't. About a year after I arrived in Canada I got married and had two children. Unfortunately, the hectic pace of my academic career resulted in me raising them as a single parent. They lived with me as I was climbing the academic ladder, and I'm proud to tell you that they have both grown into fine adults, living productive, healthy lives. I'm also proud of the fact that my ex-wife and I are good friends. The marriage may not have lasted, but the good we brought into the world endures. I eventually remarried, and my family has grown with the inclusion of my third child, our youngest daughter.

This is from an article on the development of evidence-based medicine:

> In the spring of 1990 the young McMasters University Internal medicine residency coordinator, Dr. Gordon Guyatt, had just introduced a new concept he called "Scientific Medicine." The term described a novel method of teaching medicine at the bedside. It was built on groundwork laid by his mentor Dr. David Sackett, using critical appraisal techniques applicable to the bedside. However, the response from his fellow staff was anything but warm and inviting. The implication that current clinical decisions were less than scientific, although probably true, was nonetheless unacceptable to them. Guyatt then returned with a new title that described the core curriculum of the residency program: "Evidence-Based Medicine" (EBM).[30]

I think it's important to spend some time reiterating and talking about how McMaster University, where I received my doctorate, is where the field of evidence-based medicine was created.[31] Although no official rankings exist of that program, it has been my experience that experts in the field rate it even above the programs at Harvard and Stanford. I also believe it's important for the average person to understand how fragmented the medical education is of the typical doctor. First of all, one must understand the incredible time demands placed upon the typical physician. There's meeting with patients and talking to them (the most important thing, in my opinion), then the hassle of dealing with insurance (often the thing upon which they spend the most time), as well as meeting with colleagues and the difficulty of trying to run a profitable practice. Little time is left in the day for catching up on the latest medical findings, as well as taking the hours necessary to understand how that newly published research may agree with or contradict previous medical publications. In actual practice, most physicians are forced to take the short-cut of relying on anecdotal findings from their colleagues and what they might have learned years ago, or a quick and superficial review of an article that might be getting a good deal of coverage in the media.

What we try to do in evidence-based medicine is take a more rigorous, scholarly approach. We make sure we are looking at the entire body of knowledge and evidence in a certain area, and then critically appraise that body of knowledge, because not all studies are of the same quality. You set the low-quality studies aside, noting that they may be true but their reliability is questionable, and place above them the high-quality studies of which you can be more confident in their findings. In this area, we look at the estimates of effect that are published and we appraise the robustness, rigor, and trustworthiness of the underlying evidence. In evidence-based medicine, we have established a certain set of tools to help us in that process, such as systematic review, meta-analysis, and systematic analysis, so we understand how all the pieces fit together. This type of research is done in a very systematic, open, and transparent manner, which is easily reproducible for another researcher in evidence-based medicine or within the field of academic or clinical research.

I always felt I was on the side of the physician who was trying to make the very best decision for his or her patient. Doctors and patients deserve to have the most up-to-date body of evidence possible. I also thought we were creating a record so that any researcher who had questions about a particular course of therapy could easily access the information and follow our thinking. In other words, I wanted what I published to be reproducible. And if there was a weakness in your research or your analysis, another scientist could follow what you had done, and then say to themselves, "Here is the problem. This wasn't clearly analyzed. This is an important point they didn't fully understand."

Evidence-based medicine is a standardized, open, transparent, and explicit way for the academic research community to connect directly to practitioners who are seeing patients. In addition, evidence-based researchers are the ones who are perfectly positioned to talk to politicians making public health decisions and policy, especially in the midst of a global pandemic.

Much of what passes for medicine today is what I call, "medicine out of a black box." What do I mean by this? Some research is done, maybe some good results are obtained, and before you know it, the results are reported, and the therapy is being offered to patients. What was the process by which the therapy was judged to be safe and effective? Doctors are just like everybody else. Sometimes they simply start doing something because all the other doctors are doing it. That's not the way to provide health care.

We as evidence-based researchers and methodologists want to see what is inside the black box and, in opening the black box of research and clinical medicine, we ask researchers and clinicians to lay out their evidence, read up on the latest research, and talk about it in an open and transparent manner, and then we see if the therapy or research is based on firm, trustworthy, reproducible evidence. What was often found, though, was an absence of strong evidence and instead the presence of habit. Doctors had started doing something but had never taken a rigorous look as to whether what they were doing was actually helping. They just did what other doctors did and what they heard. Dr. Guyatt and Dr. Sackett wanted much more explicitness and transparency in clinical medicine and thus the field was born in Hamilton, Ontario.

The reality is that nothing in science is ever supposed to happen in a black box.

I finished my doctorate at McMaster in 2015, did a post-doc for a year, and then was asked to join the faculty as an assistant professor in evidence-based medicine and epidemiology. In that position, I was teaching students and doctors in both the masters and doctoral programs. The doctors usually had their own private practices or were working for major hospitals, but they were taking these additional classes because they wanted to provide the very best care for their patients, and some of them wanted to learn how to conduct basic academic research.

I was lecturing in epidemiology at McMaster and was also asked to work with the Infectious Disease Society of America (IDSA), headquartered in Richmond, Virginia, to review and help develop their clinical practice guidelines on the use of drugs and other therapeutics for an infection. My work was so highly regarded that, in 2019, I was offered a position at the World Health Organization (WHO) in Geneva, Switzerland, as well as the Pan American Health Organization (PAHO), which was headquartered in Washington, DC. I was a highly respected and sought-after expert in the use of evidence-based medicine and research methodology within the field of academic research and clinical medicine.

Around December 2019, I began to hear from my sources about a possible viral outbreak in China. In January 2020, as the first reports of this virus started making their way into the media, the WHO, for whom I was already working, asked me to become their pandemic advisor, as an evidence-based synthesis advisor. Yes, the WHO and PAHO asked me to become their senior pandemic advisor.

I hope it's clear to you now, when the attacks were launched against me—that I wanted children to die and that I was some scientist who was not to be trusted—that they were the furthest things from the truth. But I'm getting ahead of myself.

How did I react to being asked by the WHO and PAHO to be their consultant and pandemic advisor?

At first, I was shocked. But as I looked at the situation, I could understand why I'd been asked to accept such a prestigious position. I dealt with evidence. The coronavirus was a new thing in the world, and the response should be driven by the very best evidence. Those were the skills I possessed in terms of being able to gather the evidence, critically appraise it, and analyze and synthesize it for optimal reporting.

I realized this was exactly the situation that called for evidence-based medicine. In the first few months of the pandemic, it was me behind the scenes at the WHO and PAHO who was providing guidance to my directors and who then gave that information to nations around the globe. I simply fell back on my training to collect the evidence, synthesize it, and then communicate that information to the world.

In early May 2020, I got a call from Washington that the White House had been reading what I'd written and spoken about and wanted me to work behind the scenes to make sure the recommendations of the COVID-19 task force were consistent with evidence-based medicine. I was informed that they wanted someone with my skills who could be trusted, as well as who wanted to work for the United States.

I was told I'd have a direct line to the secretary of Health and Human Services (HHS), Alex Azar, as well as the assistant secretary, and that we all would report directly (and at times indirectly) to the White House. Officially, though, I'd be an employee of HHS, which would become an important factor in my eventual fate. Even though I was a Canadian and the United States was completely shut down, I was told that if I went to the border at Niagara Falls, I would be met by authorized persons who would bring me across.

However, what they didn't realize was that I'd already received a green card of residency about a year earlier, based on my niche skills in evidence-based medicine and research methods. The US Citizenship and Immigration Services assessed my background and expertise and determined that I possessed skills and expertise that would benefit the United States. There is a very niche category in the US immigration system where persons with specialized skills needed in the United States are strongly considered for residency. I'm not sure native-born Americans realize how unusual it is for a foreigner to be granted a green card so quickly without needing to wait for agonizing years. But it's simply an indication of the high regard in which I was held for my expertise in evidence-based medicine and its application to infectious diseases, that I was granted an expedited residency green card.

At the time it happened I reflected how, as a young man, it had been my great desire to become an American, and here I was a man in middle age, finally achieving this goal. I didn't need the United States military to cross the border. I could come on my own, which I did.

Perhaps you will laugh at my idealism, but when I arrived in Washington, DC, one of the first calls I made was to the United States military. I told them I wanted to enlist. The recruiter patiently explained to me that the cut-off for enlistment was thirty-nine years old, and since I was fifty-four at the time, I'd missed my chance by fifteen years. He added, however, that in my position I was serving the United States just as much as any member of the military.

I had finally made it to America, the land I'd dreamed about since I was a young man. And instead of defending her in a fighter jet, I'd be part of the COVID-19 task force at HHS, fighting the greatest public health crisis of the new century.

I again thought of Dr. Eric Williams, the first prime minister of a free Trinidad and Tobago, who told us that our future lay in our school bags and that when we went out into the world, we should carry our island on our sleeve and believe that anything was within our grasp. And that we bow down to no one.

The era of the slave master was over.

However, I was to learn of a greater slavery being planned, where the citizens of the entire world would cower in the shadow of Dr. Fauci's edicts, which were not grounded in science. As far as I could determine, Dr. Fauci's "advice" seemed to be based only on how much they might damage the reelection chances of President Trump. This became clear to me quickly.

But that was in the future.

What was close at hand was the approach of other forces, not those of Dr. Fauci, but of genuine patriots, who understood the groundwork which was being laid and hoped to fight it, using me as their sword and shield.

I was the soldier they had chosen to spark the slave rebellion.

CHAPTER TWO

The Secret Meeting

The second day I was on the job at HHS in support of the COVID-19 task force, I was asked to attend an important meeting at a restaurant near the Senate and the National Mall. Although I was working in the middle of a global emergency, I still felt an enormous appreciation to be in the capitol city of the United States, sitting in a restaurant looking out on that great enormous expanse of grass, viewing the Washington Monument, the State Department, Senate, and Supreme Court just down the street. My wife decided to attend the meeting with me.

Our table was outside, which was wonderful for appreciating the monuments, but it was a hot, late spring day and I could feel myself perspiring. Washington can get as muggy as any Southern state swamp. I did not know who I was meeting or the meeting's purpose, but I decided to simply go with it.

Three people approached, wearing masks, and took them off as they sat. I took a deep breath. The first was a Republican senator, the second was a Democratic senator, and the third was a Democratic congressperson, as introduced. I was asked to keep the meeting highly confidential and secret. Considering all that has happened, while still honoring the spirit of our discussions, I am keeping the names of the politicians who met with me that day anonymous.

If they choose to identify themselves later, that is their choice.

"Hello, Dr. Alexander, we're all very excited to meet you," said the Republican senator, opening the conversation.

I'm usually very talkative, but I was shocked into silence.

"Paul, you know that Trump appointed you, correct?" the Republican senator continued.

"Y-Yes," I managed to stammer. At that time, I was going through duress at HHS with the bureaucracy that was refusing to complete my hire as well as complete my pay. I knew I was working at HHS but was unsure if I was a political appointee or an actual federal employee. Yet what I did know was that I came to Washington on request of the Trump administration.

He continued, "And you know it's because he respects your background in evidence-based medicine, right?"

"Yes," I replied more confidently.

"That's only part of the objective," said the Republican senator.

"What's the other part?" I asked, uncertain where this conversation could possibly be leading.

"We have an objective," said the Democratic senator, speaking for the first time. "Our objective is to take down the CDC. Take it down to the studs. Complete. We want to remake it from the ground up, not destroy it, but to fix it, to remake it to the premier agency it should be. But first we have to investigate it, dismantle and restructure it."

I looked at my wife. As much as I'm a talker, she's the one I confide in, the person whose counsel I value the most. Her eyes were as wide with surprise as mine. I looked back to the Democratic senator. "Where do I come in?"

"Well, we know your skills and talents," said the Democratic senator, motioning to the two politicians accompanying him, who both nodded. "We actually had a role in you coming here and asked certain people to find you. Yes, it's true you were hired by the Trump administration, but we're all part of this."

"And what is this?" I asked.

The Democratic congressperson picked up the thread. "In our opinion, and in the opinion of many in congress, the senate, and in many branches of government, we've all come to the conclusion that the CDC is now operating as a political agency. It's been corrupted by Big Pharma money. That's what's driving the COVID-19 response, not the science. Just how it's going to benefit the pharmaceutical industry. We can't have that. It is hurting the president's response to the pandemic because every time he tries to act, he gets undercut. If it were another president in power, our conclusion would be the same. Again, this is not only our view and we might add, the same is felt for other health agencies such as the NIH and FDA. The president can't act, even when the science is on his side, because the CDC is against him. He doesn't have the ability to evaluate the CDC's reports and statements, so we need somebody to help him with it. We need you to help us take down the CDC because they are wrecking America, the country we all love."

"I'm just one person, and I still do not understand what you are asking of me," I protested. I also reminded them that "my understanding of my COVID advisory role at HHS was that of a primary and acute focus on the coronavirus pandemic [anything asked of me at HHS as well as the White House] as well as a broader array of issues in my portfolio related to COVID and could be requested by my manager as they deemed fit." Yet I now understood that my role really was this one acutely focused on the CDC. I was not told this before I arrived in DC.

The Democratic congressperson continued, "We know your background in evidence-based medicine. We know you did not know this is why you are principally here. We need somebody like you firstly that is trustworthy and we have concluded this. Your role would be to go into the CDC's work [with even comparisons to the 2009 swine flu pandemic under President Obama and how the CDC and other health agencies reacted and the reports they put out] and review it because we don't have that skill set. No one in the senate or congress who has oversight does. We've talked to people and it's clear to us you love America and want to help in this pandemic. We have heard you speak internationally and actually read your writings. It is clear you are overwhelmingly qualified. We did a lot of research on you before the reach-out and job offer was made to you. We even know how you got your expedited green card because there are so few in the United States who have your knowledge and experience. We're asking for your help, Paul, and we will ensure that all of the right clearances are in place as you proceed."

I took a moment to process what they were telling me. "Okay," I said, "some kind of official report for you to examine?"

Yes," said the Republican senator, taking control of the meeting. From the way the other two were treating him, it seemed he was the head of this delegation, although they were both enthusiastic participants. "And we want you to write the report quickly because we want to put it in front of congress and the senate so there can be a high-level oversight discussion of what's going on at the CDC. Your reputation as a scientist specializing in evidence-based medicine and research will make it impossible to ignore. That will force a hearing. We will bring in everybody to answer to that report. We want you to go into all of the CDC's statements, analyses, reports, as far back as you can but let us focus too on the COVID response, and show us everything they've done wrong. Anything questionable. Write it out for us. Everything they've done wrong. We need to get this out in front of people.

We're already in the third month of people's lives being uprooted. This has to stop."

"Is this a request to write this report? A suggestion?"

"No," said the Republican senator. "This is a direct request and it's coming from the White House. The Oval Office."

I looked at them again. "Is this a real request?" I asked.

"Yes," they replied. "This is a real request, and this has to be done very secretly. Nobody is to know until the report is done. Also, we want such a report by October 1."

I was getting a little more comfortable with the idea of analyzing a bunch of evidence and writing a report. After all, that's pretty much what I did in my evidence-based medicine career. I reviewed publications, data, and practices, then wrote a critique. This would simply be more of the same thing.

Except for the fact it was intended to demolish and remodel the CDC down to its studs. To fix it. This appealed to me, as well as the potential similar approach to the other health agencies.

"Okay, who is going to be on my team?" I asked.

"Nobody else. Just you," said the Republican senator.

"How can I be the team?"

"You have the ability."

I was being confronted with an impossible task. "Yes, I know I have the ability. But that's not the question. This is something I can manage, but not do all on my own. I need people doing the work. Checking it and rechecking it. I can then come in and finesse it, make sure it's bulletproof. But I can't do all of this on my own."

The Republican senator was adamant. "That's the only way this can be done. Extreme security and this will ensure no leaks, etc. before completion, for if the press gets wind, they will embark on a smear campaign of you and your credentials. There will be one name on the report and that's yours. This is DC and if you have a team, there will be leaks. We can't have that. This has to be something that will surprise everybody when we bring it out."

"Only my name on the report?" I asked, and then glanced over to my wife.

She was shaking her head. I knew that meant no. And I trusted her in these situations.

The Republican senator wanted to give me one more pitch before the group left. "Paul, we need your help. This would impact funding for the CDC, giving us the ability to remodel and remake it into something that actually serves the people. The CDC has become too political and biased, and they are making terrible errors and mistakes in their guidance. But, we need to strip it down first, so people can plainly see what they've been doing. And in order to do that we need to have something in our hands to show the American people, so they understand the corruption of the CDC. We have been trying to do this silently, but we can't. Somebody told us you are that kind of silent person we could trust, and that you love this country.

"But you should know from us, this is Trump, your president, directly asking you to do this. He knows the failures across the health agencies and especially what has gone on thus far with the pandemic response. As the president, we know he wants the agencies to improve their functioning, their credibility, and their output guidance to the American people that, as is shown now, is often almost always wrong and without any scientific basis. We feel strongly that any president must have the best health agencies like CDC and FDA reporting to him or her and at present, this president does not, and we want to fix that for him and any future president."

The group got up to leave, each one shaking my hand, and saying they understood I was in a difficult situation and hadn't really known what I had signed on for. My mind was racing at that point, and I was genuinely conflicted. I understood that the CDC was a corrupt, inept, bogus political agency, which probably deserved to be demolished, remade, and restructured properly, but why did I have to be the single gladiator in the arena?

Later, when my wife and I got to our apartment near the Capitol Building, she shared her fears. "Paul, you can't do this," she pleaded. "They will kill you. They will put a bullet in your head. And why is it that they only want your name on the report? Whoever these people are, they'll get what they need from you for their purposes, but you won't survive. Please don't do this. This is way bigger than you realize."

I assured her I'd be safe, but my mind kept coming back to the request. As a young man I'd wanted more than anything to be like Tom Cruise in *Top Gun*, the "best of the best," putting his life on the line for the greatest country in the world. I loved all that the United States stood for as really the last beacon of hope in the free world. If there is ever to be real democracy, I knew it was where I was now living.

Now as a grown man I'd been handed that very opportunity.

There was so much work to be done at HHS in my supportive role on the COVID-19 task force and communications team that it would take up almost all my waking hours.

As much as I shared my wife's suspicions of the true motives of the three politicians, I kept asking myself, *What if everything is just as it's been presented? What if good and patriotic Americans of all political stripes were understanding the peril of a Big Pharma corrupted CDC (and NIH and FDA and NIAID) and wanted to remove that threat?*

I genuinely wanted to believe there were good people working in the United States government, who had seen what had gone wrong and desperately wanted to fix it. And it made sense that they would try to bring in somebody with my set of skills in order to assist them. I was really amazed and inspired that there were people like the congressperson and senators I had met. They seemed to love the nation and wanted to do good.

Despite the crushing work requirements of HHS and supporting the COVID-19 task force, the fact that I was completely on my own, and my wife's fear that my career would be ruined or I might be murdered, I began working on the report the three politicians hoped would demolish the CDC down to its studs, and help remake it into the marquee agency it should have been.

I was all in.

* * *

How do you keep yourself focused on tackling the largest public health crisis of the past century while at the same time holding in your brain the idea that your real purpose is to prepare the report that is intended to bring down the CDC, which was probably the most trusted agency in the United States government?

Some people hated the politicians, questioned the intelligence agencies, or believed the police were racist and needed to be defunded, but ironically, they still trusted their scientists and doctors. During the pandemic, images of Dr. Fauci would be sold on candles as if he was a Catholic saint by people who didn't believe in religion in the first place. Dr. Fauci would even be played by Brad Pitt in a *Saturday Night Live* segment, for which Pitt would receive an Emmy nomination.[32]

The way I handled the undertaking was the same way I handled many of the most challenging problems in my life: compartmentalization.

Yes, I might feel the weight of the world on my shoulders, but I simply put one step in front of the other. The alarm rings, I get up. I go to the bathroom and take my shower. Get dressed, go to the kitchen to get my coffee, visit with my wife and youngest daughter, listen to their problems and concerns, then drive to my office on the sixth floor of HHS and confront the challenges that are waiting for me.

One of the things that concerned me at the outset of my time working at HHS in support of the COVID-19 task force was trying to figure out the origin of the coronavirus and, perhaps more importantly, why all the measures that were being put forth were such a stark departure from decades of accumulated wisdom on dealing with viral outbreaks.

Dr. Deborah Birx would later say that SARS-CoV-2 "came out of the box, ready to infect" and that it was either some freak of nature with capabilities never seen or was designed in a lab.[33] I was conflicted by a similar question, since when I consulted the evidence it seemed the public health officials at WHO and the Infectious Disease Society of America had either:

1. Made one of the greatest predictions in human history regarding a viral outbreak, or,
2. Planned and executed a deadly, worldwide viral outbreak.

The evidence could be read to support both possibilities, and with similar probabilities, meaning that if I was biased in the slightest amount in either direction, I would read the evidence in exactly the way I wanted to read it. This is from a CNN article from September 18, 2019, with the title, "The Risk of a Global Pandemic is Growing—And the World isn't Ready, Experts Say":

> The chances of a global pandemic are growing—and we are all dangerously under prepared, according to a new report published Wednesday.
>
> The panel of international health experts and officials pointed to the 1918 influenza pandemic as an example of a global catastrophe. That killed as many as 50 million people —if a similar contagion happened today, it could kill up to 80 million people and wipe out 5% of the global economy.
>
> "The world is not prepared," the report from the Global Preparedness Monitoring Board (GPMB), co-convened by the World Bank and the World Health Organization (WHO), warned. "For too long, we have allowed a cycle of panic and neglect when it comes to pandemics: we ramp up efforts when there is a serious threat, then quickly forget about them when the threat subsides. It is well past time to act."[34]

What was really going on with this article? Was it simply trying to prepare us for a possible outbreak, or was it, to use the words of the more suspicious in our society, "predictive programming" designed to prepare us for some plan formulated by individuals who would forever remain in the shadows?

It's said that the best way to fool people is to let them in on a small part on your lie, and later act as if you hadn't said it. Then, when you're surprised by the turn of events, the offending party will turn around and say, "I told you exactly what I was going to do. If you had an objection, why didn't you say something at the time?"

How else can one explain the following two paragraphs?

> Poorer countries, especially those without basic primary health care or health infrastructure, are hit hardest by disease outbreaks. In these places, the problem is often compounded by armed conflict or deep distrust in health services, as seen in the Democratic Republic of the Congo (DRC), which has been ravaged by an Ebola outbreak for more than a year. Community mistrust has led to violent, sometimes fatal attacks on health care workers.
>
> Scientific and technological advancements have helped fight these diseases—but the WHO report warns they can also provide the laboratory environments for new disease-causing microorganisms to be created, increasing the risk of a future global pandemic.[35]

In one paragraph we're informed about an Ebola outbreak in the Democratic Republic of Congo, as well as the fact that the locals don't trust health care workers. In the next we're told that scientists in labs are capable of creating "new-disease causing microorganisms." In other words, they tell us that ignorant Africans are suspicious of Western health care workers. And in the next they tell us that the same Western medical system might be creating new diseases.

I'm having trouble figuring out who the ignorant ones are in this scenario.

And if we go back to the earlier part of the article, where they mentioned the 1918 global influenza pandemic that killed fifty million people, you'd probably be saying to yourself, "Well, at least I know scientists aren't stupid enough to mess around with that virus!"

But you'd be wrong.

This is from the CDC's website under the title, "Reconstruction of the 1918 Influenza Pandemic Virus":

> CDC researchers and their colleagues successfully reconstructed the influenza virus that caused the 1918-1919 flu pandemic, which killed as many as

50 million people worldwide. A report of their work, "Characterization of the Reconstructed 1918 Spanish Influenza Pandemic Virus," was published in the October 7, 2005 issue of *Science*. The work was a collaboration among scientists from CDC, Mount Sinai School of Medicine, the Armed Forces Institute of Pathology, and Southeast Poultry Laboratory.[36]

Yeah, so you may not have known it, but in 2005 scientists from the CDC recreated the deadliest pathogen of the past hundred years. And it wasn't just the CDC; it was also the Armed Forces Institute of Pathology.

That's right—the military was involved in this effort.

A little further into the website it helpfully tells the public that the work was also sponsored by the "U.S. Department of Agriculture," and the "National Institutes of Health."[37]

In the section titled "Biosecurity Issues," this was the first "question" and subsequent "answer" designed to make you feel better about scientists playing around with a virus that had killed fifty million people:

Did the generation of the 1918 Spanish influenza pandemic virus containing the complete coding sequence of the eight viral gene segments violate the Biological Weapons Convention?

No. Article I of the Biological Weapons Convention (BWC) specifically allows for microbiological research for prophylactic, protective, or other peaceful purposes. Article X of the BWC encourages the "fullest possible exchange of . . . scientific and technological information" for the use of biologic agents for the prevention of disease and other peaceful purposes. Further, Article X of the BWC provides that the BWC should not hamper technological developments in the field of peaceful bacteriological activities. Because the emergence of another pandemic virus is considered likely, if not inevitable, characterization of the 1918 virus may enable us to recognize the potential threat posed by new influenza virus strains, and it will shed light on the prophylactic and therapeutic countermeasures that will be needed to control these pandemic viruses.[38]

This answer is full of half-truths and evasions. If you're to believe the text of this answer, it's the "intention" of the person doing the research that determines whether it's allowed or banned.

Here's the problem.

How can one possibly know the intention of everybody working on the research? What if most of the researchers have good intentions, but some from the "Armed Forces Institute of Pathology" are thinking to themselves, "Hey, I found myself a great new weapon!"

And in returning to the CNN article from September 2019, I think it's instructive to look at the report they mentioned, as it has many interesting sections. It's put together by the Global Preparedness Monitoring Board, and curiously has a picture on the front of a large group of people all wearing masks.

Predictive programming?

I'll let you be the judge.

As with any political campaign, it always helps if you know the donors. However, they're a little vague on that when they write:

> We would like to express our deepest gratitude to the Board Members and the staff who gave their time, wisdom, and contributions to shaping this report. This report would not have been possible without the many individuals from academic institutions, experts, multilateral agencies, non-governmental organizations, and national governments who willingly gave their time, insights, experiences and contributions to the Board, to the Secretariat, and especially the teams developing the background papers for the GPMB.[39]

It doesn't really read like much of an acknowledgments section. We've got "academic institutions, experts, multilateral agencies, non-governmental organizations, and national governments."

But that's it.

Surely these groups wanted to be recognized. Everybody wants a thank you, don't they? And the biggest contributors most surely receive their thanks at the end where the authors write: "Finally, we are grateful for the financial support provided by the GPMB Secretariat from the Government of Germany, the Bill and Melinda Gates Foundation, the Wellcome Trust, and Resolve to Save Lives."[40]

You just knew that eventually we'd run across Bill and Melinda Gates in all of this, didn't you?

Let's dive into some of the report's plans for the citizens of the world, shall we? Here are their seven action items:

1.   Heads of governments must commit and invest.
2.   Countries and regional organizations must lead by example.
3.   All countries must build strong systems.

4. Countries, donors, and multilateral institutions must prepare for the worst.
5. Financing institutions must link preparedness with financial risk planning.
6. Development assistance funders must create incentives and increase funding for preparedness.
7. The United Nations must strengthen coordination mechanisms.[41]

That's an interesting list, compiled and published just three months before this novel virus escaped (I mean emerged!) from some still unknown location in Wuhan, China, which just by coincidence had both a wet market and a Biosafety Level 4 Lab that was working on bat coronaviruses.

For each one of these "urgent action calls" there were one to three "progress indicators"; probably none were more interesting than the first "progress indicator" for the third "urgent action call," which was:

> At a minimum, the 59 countries that have completed a NAPHS [National Action Plan for Health Security] identify a national high-level coordinator (board, commission, agency) to implement national preparedness measures across all sectors, and to lead and direct actions in these sectors in the event of a public health emergency.[42]

And when you went to the back of the report, you found among the fifteen members of the Global Preparedness Monitoring Board Dr. Chris Elias, president of the global development program of the Bill & Melinda Gates Foundation.[43]

It's almost as if you got all the devils together in one place just before the plague descended on the world. If you think that maybe I'm being a little too alarmist, tell me how you interpret the following section:

> **The chances of a global pandemic are growing.** While scientific and technological developments provide new tools that advance public health (including safely assessing medical countermeasures), they also allow for disease-causing microorganisms to be engineered or recreated in laboratories. A deliberate release would complicate outbreak response; **in addition to the need to decide how to counter the pathogen, security measures would come into play limiting information-sharing and fomenting social divisions**. [Bold by authors.] Taken together, naturally occurring, accidental, or

deliberate events caused by high-impact respiratory pathogens pose "global
catastrophic risks."[44]

One would have to twist the language to interpret it as anything other than
a plan to lie to the public in the case of what they say is a "deliberate release."
Left unanswered is the question of whether this same protocol would be
in place during an "accidental release." I don't think it's stretching the language at all to say that, in the case of a deliberate release, they were planning
an active misinformation campaign to cover up the truth, even knowing
their campaign of lies would not be believed by a significant number of the
public and would cause social unrest.

I really must spend some time discussing the blockbuster nature of this
passage. Science is supposed to be about truth, right?

Follow the science.

That's what they told us during the COVID pandemic.

And we were supposed to do that because the one value science supposedly held above all others was the relentless search for the objective facts.

And yet, this approach was a glaring exception.

Tell the truth, they were saying, if it was a natural outbreak.

But if scientists caused the outbreak (because they like to play with dangerous pathogens in our labs) and lied their asses off, even when challenged,
maybe our best play is to claim those telling the truth are "conspiracy theo-
rists." Is it any wonder that one of the following paragraphs noted:

> **Trust in institutions is eroding.** Governments, scientists, the media, public
> health systems and health workers in many countries are facing a breakdown
> in public trust that is threatening their ability to function effectively. The
> situation is exacerbated by misinformation that can hinder disease control
> communicated quickly and widely via social media.[45]

It's difficult to overstate how clueless these scientists sound to the general
public. They genuinely seem to believe that they are smarter and know better than the average man or woman. There's a saying that "there are some
things so stupid (like communism or socialism) that only an academic can
believe" and that might just be the case here. Consider the argument they
make:

They want the ability to experiment on dangerous pathogens in their
labs.

If something goes wrong, and one escapes, they want to have the ability to lie to the public about it.

And when a significant number of the public doesn't believe their lies, they want to have a way to isolate and shame us.

And these same scientists want us to believe that's a great way to run society, and we'll all be so much happier with their benevolent rule over us.

In the "Progress to Date" section of the report, they let you know how far they've come in their plans to have a system of control in place for any possible pandemic outbreak, whether natural or artificial.

> In 2017 Germany, India, Japan, Norway, the Bill and Melinda Gates Foundation, the Wellcome Trust and the World Economic Forum founded the Coalition for Epidemic Preparedness Innovations (CEPI) to facilitate focused support for vaccine development to combat major health epidemic/pandemic threats.[46]

How well-structured was the media campaign between the various nations, experts, and "charitable foundations" in advance of the COVID-19 outbreak? You had probably the most important country in Europe, Germany, on board, as well as the more than a billion people of India, the economic powerhouse of Asia, Japan, as well as Dr. Anthony Fauci, Bill and Melinda Gates (pre-divorce, before Melinda fully realized the horror of her husband's association with billionaire pedophile Jeffrey Epstein)[47], and Klaus Schwab of the World Economic Forum.

Kind of sounds like an all-star team who shouldn't be afraid of guys in their underwear in their mother's basement reading and analyzing research papers, then publishing what they found on the internet, right?

* * *

What did the real experts have to say about these issues?

I previously mentioned Dr. Donald Henderson, my former professor at Johns Hopkins University, who led the campaign to eradicate smallpox and was a "bioterrorism advisor to several presidents." At the age of seventy-eight, in the wake of the SARS-CoV-1 outbreak, Henderson published a seminal paper titled "Disease Mitigation Measures in the Control of Pandemic Influenza." (Since the world has had far more experience with influenza viruses than coronaviruses, influenza mitigation efforts have been looked upon as a model for coronavirus outbreaks.) In the 2006 paper, Henderson stated the problem of how to respond to a pandemic:

> Possible measures that have been proposed include: isolation of sick people in hospitals or at home, use of antiviral medications, hand-washing and respiratory etiquette, large-scale or home quarantine of people believed to have been exposed, travel restrictions, prohibition of social gatherings, school closures, maintaining personal distance, and the use of masks. Thus, we must ask whether any or all of the proposed measures are epidemiologically sound, logistically feasible, and politically viable. It is also critically important to consider possible secondary social and economic impacts of various mitigation efforts.
>
> Over the years, various combinations of these measures have been used under epidemic and pandemic circumstances in attempts to control the spread of influenza. However, there are few studies that shed light on the relative effectiveness of these measures.[48]

All those measures you thought were new for COVID-19? They'd been discussed for years in academic journals. Let's look at the worst-case scenario put forth by Henderson and his fellow authors.

> Many communitywide disease mitigation measures would be intrinsically difficult to implement. Consideration must be given to the resources required for implementation, to the mechanisms needed to persuade the public to comply (or to compel the public, if the measures are mandatory), and to the length of time that they would need to be applied. Potential disease mitigation measures presumably would have to be maintained for the duration of the epidemic in a community—a predicted period of 8 or more weeks—or, perhaps, in the country as a whole—as long as 8 months.[49]

So, we were looking at anywhere from eight weeks to eight months. Why was our lockdown so much longer, with students going back to school more than a year after the nationwide school closures in March 2020?

Was it because of a little something called the presidential elections in November 2020? If the schools were closed, businesses shuttered, and people still in a state of panic, it was much easier to loosen those voting laws for the benefit of the opposing political party, in this case the Democratic Party.

In summation of all these possible responses, it was the opinion of the authors that only two measures held any real promise in the event of a novel pathogen, and those were the isolation of sick individuals and increased handwashing. The authors had harsh words for the idea of quarantining healthy people and the inevitable social discord this would provoke.

> As experience shows, there is no basis for recommending quarantine either of groups or individuals. The problems in implementing such measures are formidable, and secondary effects of absenteeism and community disruption as well as possible adverse consequences, such as loss of public trust in government and stigmatization of quarantined people and groups, are likely to be considerable.[50]

That thing we did for nearly two years? The lockdowns? Not supported by the science. This is from the scientist who led the campaign to eradicate smallpox. This was the summation of their findings:

> Experience has shown that communities faced with epidemics or other adverse events respond best and with the least anxiety when the normal social functioning of the community is least disrupted. Strong political and public health leadership to provide reassurance and to ensure the needed medical care services are provided are critical elements. If either is seen to be less than optimal, a manageable epidemic could move toward catastrophe.[51]

Did Dr. Anthony Fauci and the COVID-19 task force respond by lowering people's anxiety, or cranking it up to the equivalent of a four-alarm fire? You know the answer to that one. Could we say that a "manageable epidemic" became a "catastrophe"? That's how I'd describe it.

But maybe a paper from 2006, even by one of the giants in the field of public health, escaped Dr. Fauci's attention and that of the COVID-19 task force.

* * *

However, what about a report by the World Health Organization (WHO) in 2019 titled "Non-Pharmaceutical Public Health Measures for Mitigating the Risk and Impact of Epidemic and Pandemic Influenza"? That sounds like pretty much the first paper the public health officials should have consulted with COVID-19. This WHO report was compiled by scientists from many global nations and thus one can say that the global community, in

2019, had a blueprint for how it should and must respond in epidemics and pandemics. This is from the Introduction to the Executive Summary:

> Influenza pandemics occur at unpredictable intervals and cause considerable morbidity and mortality. Influenza virus is readily transmissible from person to person, mainly during close contact, and is challenging to control. In the early stage of influenza epidemics and pandemics, there may be delay in the availability of specific vaccines and limited supply of antiviral drugs. Non-pharmaceutical interventions (NPIs) are the only set of pandemic countermeasures that are readily available at all times and in all countries. The potential impacts of NPIs on an influenza epidemic or pandemic are to delay the height and peak of the epidemic if the epidemic has started; reduce transmission by personal protective or environmental measures; and reduce the total number of infections and hence the total number of severe cases.[52]

Non-pharmaceutical interventions are those interventions which are immediately available in all countries. Therefore, they should be seen as the first line of defense.

So, what did they consider and how effective did they believe these measures would be?

There were several possible measures considered, and they were rated for the quality of the data regarding their effectiveness, the strength of the recommendation, and when they should be followed, depending upon how much of a challenge it would be to adopt the measure. In summary, they were:

1. Handwashing—Moderate evidence it did not work but recommended at all times since it was simple to implement and already part of people's lives.
2. Respiratory etiquette—No evidence of effectiveness but recommended at all times.
3. Installing UV lights in enclosed and crowded places—No evidence and not recommended.
4. Increasing ventilation to reduce transmission—Very little evidence (but some) it was effective, but recommended at all times.
5. Modifying humidity—No evidence and not recommended.
6. Active contact tracing—Little evidence of effectiveness and not recommended.

7. <u>Isolation of sick individuals</u>—Very low evidence (but some) it was effective, but recommended at all times.

8. <u>Quarantine of exposed individuals</u>—Very low amount of evidence, and variable effectiveness. Not recommended.

9. <u>School measures and closures</u>—Very low amount of evidence, but some showing effectiveness. May do partial closings, or full in severe epidemic or pandemic.

10. <u>Workplace Measures and Closures</u>—Very low amount of evidence (but some) showing effectiveness. Conditionally recommended in extraordinarily severe epidemics and pandemics.

11. <u>Avoiding crowding</u>—Very little evidence, can't determine effectiveness, but would conditionally recommend in moderate and severe epidemics and pandemics.

12. <u>Travel advice</u>—No evidence but recommended during the early phase of a pandemic.

13. <u>Entry and exit screening</u>—Very low amount of evidence and tending to show a lack of effectiveness in reducing transmission. Not recommended.

14. <u>Internal travel restrictions</u>—Very low amount of evidence, but some showing of effectiveness. Conditionally recommended in the early phase of an extraordinarily severe pandemic.

15. <u>Border closure</u>—Very low amount of evidence, and evidence of effectiveness is mixed. Not recommended unless required by national law in extraordinary circumstances during a severe pandemic.

16. <u>Face masks</u>—Moderate amount of evidence on question of effectiveness but tending to show they were not effective at stopping transmission. Recommended for symptomatic individuals, but conditionally recommended for asymptomatic individuals in a severe pandemic.

17. <u>Surface and object cleaning</u>—Good evidence that cleaning of surfaces will reduce viruses on that object, but no evidence of effect on transmission of the virus between people. However, since disinfectants are widely available, this intervention is highly feasible and recommended, as there is a "mechanistic plausibility for the potential effectiveness of this measure."[33]

How different our world would have looked if these "evidence-based" suggestions from the year before the pandemic had been followed. Citizens would have been told to increase handwashing, cover their mouths when coughing, improve ventilation, and, in severe instances, there might have been limited school or work closures (maybe they could have called it "two weeks to stop the spread") and limited travel, but otherwise life would have continued as normal.

Here's how the authors put it in summation:

> The most effective strategy to mitigate the impact of a pandemic is to reduce contacts between infected and uninfected person, thereby reducing the spread of infection, the peak demand for hospital beds, and the total number of infections, hospitalizations, and deaths.
>
> However, social distancing measures (e.g., contact tracing, isolation, quarantine, school and workplace measures and closures, and avoiding crowding) can be highly disruptive, and the cost of these measures must be weighed against their potential impact. Early assessments of the severity and likely impact of the pandemic strain will help public health authorities to determine the strength of intervention. In all influenza epidemics and pandemics, recommending that those who are ill isolate themselves at home should reduce transmission. Facilitating this should be a particular priority. In more severe pandemics, measures to increase social distancing in schools, workplaces and public areas would further reduce transmission.[54]

Isolate the sick from the healthy. We know that works.

Anything else is simply guesswork on our part. It doesn't mean we shouldn't do it. I just think it would have been better if people understood how little we actually knew about slowing or stopping a viral outbreak. Moreover, evidence quickly accumulated globally that the vast majority of the lockdown restrictive measures were ineffective and were causing far greater harm to the populations.

One of the important points I heard early on was that each day of the lockdown required ten days to recover from the economic damage. Since we were locked down for well over a year, I believe the most optimistic view is that we will be living with the economic consequences of what was done for more than a decade. There are even some economic modelers who state that it will take the rest of the twenty-first century (eighty years) for nations such as the United States to recover from the devastation of the lockdowns.

This was the final paragraph of the conclusion from the executive summary of the document looking at whether any of these measures were supported by strong evidence and showed effectiveness in reducing transmission.

The evidence base on the effectiveness of NPIs [non-pharmaceutical interventions] in community settings is limited, and the overall quality of evidence was very low for most interventions. There have been a number of high-quality randomized controlled trials (RCTs) demonstrating that personal protective measures such as hand hygiene and face masks have, at best, a small effect on influenza transmission, although higher compliance in a severe pandemic might improve effectiveness. However, there are few RCTs for other NPIs, and much of the evidence base is from observational studies and computer simulations. School closures can reduce influenza transmission but would need to be carefully timed in order to achieve mitigation objectives. Travel-related measures are unlikely to be successful in most locations because current screening tools such as thermal scanners cannot identify pre-symptomatic infections and afebrile infections, and travel restrictions and travel bans are likely to have prohibitive economic consequences.[55]

There you have it. The best evidence-based recommendations from a panel of experts convened by the WHO in the year before the COVID-19 pandemic struck our planet. These were recommendations by the very nations that turned around and hid this seminal guidance document in the bottom drawer of the desk.

Why were our public health professionals abandoning science and rationality during the COVID-19 panic? Was it because they were truly terrified?

If so, they should have been fired for their lack of judgment.

The reason they are in such positions of authority is they're not supposed to be swayed by emotion and fear, but simply the facts.

However, from what I observed, there was a different agenda at play.

* * *

Maybe all of the top medical researchers are compromised by the billions of research dollars flowing from the coffers of Dr. Anthony Fauci and Dr. Francis Collins over the years and can't be expected to render an independent judgment.

But surely there are other experts with the requisite brainpower and intelligence to crunch the numbers and examine the data to tell us what it means?

In January 2022, a study was published in the journal *Studies in Applied Economics* titled "A Literature Review and Meta-Analysis of the Effects of Lockdowns on COVID-19 Mortality" by three well-known economists, under the direction of senior author Dr. Steve H. Hanke, founder and co-director of the Johns Hopkins Institute for Applied Economics, Global Health, and the Study of Business Enterprise.

Were the three authors rigorous in their review of the data? I'll let you be the judge of their efforts.

> This systematic review and meta-analysis are designed to determine whether there is empirical evidence to support the belief that "lockdowns" reduce COVID-19 mortality. Lockdowns are defined as the imposition of at least one compulsory, non-pharmaceutical intervention (NPI). NPIs are any government mandate that directly restricts people' possibilities, such as policies that limit internal movement, close schools and businesses, and ban international travel. This study employed a systematic search and screening procedure in which 18,590 studies are identified that could potentially address the belief posed. After three levels of screening, 34 studies ultimately qualified. Of those 34 eligible studies, 24 qualified for inclusion in the meta-analysis. They were separated into three groups: lockdown stringency index studies, shelter in place order (SIPO) studies, and specific NPI studies. An analysis of each of these three groups support the conclusion that lockdowns have had little to no effect on COVID-19 mortality. More specifically, stringency index studies find that lockdowns in Europe and the United States only reduced COVID mortality by 0.2% on average. SIPOs were also ineffective, only reducing COVID-19 mortality by 2.9% on average. Specific NPI studies also find no broad-based evidence of noticeable effects on COVID-19 mortality.
>
> While this meta-analysis concludes that lockdowns have had little to no public health effects, they have imposed enormous economic and social costs where they have been adopted. In consequence, lockdown policies are ill-founded and should be rejected as a pandemic policy instrument.[56]

Do we believe this paper revealed something that was hidden from the public health professionals tasked with handling COVID-19?

No.

It was consistent with the report from the WHO in 2019 when they had looked at the question of how to handle a pandemic.

And it was consistent with the massive amount of evidence that was quickly accumulating—that already existed—with what was playing out in our lives. It was consistent with what I heard at HHS and in my capacity supporting the COVID-19 task force, being declarative on how the lockdowns were destroying lives and businesses. How the lockdowns were impacting children in terms of their emotional, social, and educational development. How many people were saved by these lockdowns? It can be difficult to estimate, especially because they are not balanced by the increase in the number of suicides, overdose deaths, domestic abuse fatalities, and undiagnosed medical problems. Some would argue that, based on the collateral damage, no one was saved by the lockdowns. They were entirely ineffective and actually harmful. But here at least is a partial answer from the report.

> With populations of 748 million and 333 million, respectively the total effect as estimated by Ashraf (2020) is 4,766 averted COVID-19 deaths in Europe and 1,969 averted COVID-19 deaths in the United States. By the end of the study period in Ashraf (2020), which is May 20, 2020, 164,600 people in Europe and 97,081 people in the United States had died of COVID-19. Hence, the 4,766 averted deaths in Europe and the 1,969 averted COVID-19 deaths in the United States corresponds to 2.8% and 2.0% of all COVID-19 deaths, respectively, with an arithmetic average of 2.4%.[57]

Each life is precious, but for comparison, in 2021, 42,915 people died in traffic fatalities,[58] and yet we keep driving our vehicles. And like I said, that doesn't take into account the additional deaths from suicide, financial stress, and domestic abuse. The authors left no doubt as to what their findings revealed:

> . . . The evidence fails to confirm that lockdowns have a significant effect in reducing COVID-19 mortality. The effect is little to none.
>
>   The use of lockdowns is a unique feature of the COVID-19 pandemic. Lockdowns have not been used to such a large extent during any of the pandemics of the past century. However, lockdowns during the initial phase of the COVID-19 pandemic have had devastating effects. They have contributed to reducing economic activity, raising unemployment, reducing schooling, causing political unrest, contributing to domestic violence, and undermining liberal democracy. These costs to society must be compared to the benefits of lockdowns, which our meta-analysis has shown are marginal at best. Such a standard benefit-cost calculation leads to a strong conclusion: **lockdowns should be rejected out of hand as a pandemic policy instrument**.[59] [Bold by authors.]

All these studies pointed to the fact that when experts looked at the actual data, they could not find strong evidence of benefit. In fact, I would argue that, if they added in the additional economic damage and lives lost from the result of unemployment, the evidence would reveal that more lives were lost than saved by the pandemic response of Dr. Fauci and his band of incompetents.

* * *

However, I guess you never would've bet that the CDC would put out their own report which repeated many of the same criticisms?

And Dr. Anthony Fauci, on his shrinking island of denial, even disagrees with the findings of his sister agency. As reported in the *Wall Street Journal*:

> U.S. states with more-restrictive policies fared no better, on average, than states with less-restrictive policies. There's still no convincing evidence that masks provided any significant benefits. When case rates throughout the pandemic are plotted on a graph, the trajectory in states with mask mandates is virtually identical to the trajectory in states without mandates. (The states without mandates actually had slightly fewer deaths per capita.) International comparisons yield similar results. A Johns Hopkins University meta-analysis of studies around the world concluded that lockdown and mask restrictions have had "little to no effect on COVID-19 mortality."[60]

The evidence now shows that lockdowns didn't matter and neither did masks. Regardless of what states did, the death rates were essentially similar. And what about schools? The *Wall Street Journal* article reviewed that as well.

> Florida and Sweden were accused of deadly folly for keeping schools and businesses open with masks, but their policies have been vindicated. In Florida the cumulative age-adjusted rate of Covid mortality is below the national average, and the rate of excess mortality is lower than in California, which endured one of the nation's strictest lockdowns and worst spikes in unemployment. Sweden's cumulative rate of excess mortality is one of the lowest in the world, and there's one particularly dismal difference between it and the rest of Europe as well as America: the number of younger adults who died not from Covid but from the effects of lockdowns.[61]

The schools and businesses should not have been closed and the masks were just hygiene theater. Those that didn't engage in this craziness had lower rates of excess mortality and fewer children who died from the lockdown. Sweden also had some other remarkable findings.

> Even in 2020, Sweden's worst year of the pandemic, the mortality rate remained normal among Swedes under 70. Meanwhile, the death rate surged among younger adults in the U.S., and a majority of them died from causes other than Covid. In Sweden, there were no excess deaths from non-Covid causes during the pandemic, but in the U.S. there have been more than 170,000 of the excess deaths.
>
> No one knows exactly how many of these deaths were caused by lock-downs, but the social disruptions, isolation, inactivity and economic havoc clearly exacted a toll. Medical treatments and screenings were delayed, and there were sharp increases in the rates of depression, anxiety, obesity, diabetes, fatal strokes and heart disease, and fatal abuse of alcohol and drugs.[62]

This paragraph contains what I believe to be some of the most shocking facts to the general population, namely that in the under-seventy population the death rate could remain unchanged, despite what was proclaimed in the media to be the worst pandemic since the Spanish Flu just after the end of World War I.

But if this virus wasn't killing those under seventy years of age (especially those without pre-existing conditions), why did the working age population, particularly those with small businesses, have to shelter at home? And how does one explain the more than 170,000 excess COVID deaths in the United States? In my estimation, that's what happens when you stress the population, both financially by ruining their businesses and emotionally by the constant drumbeat of media terror, which blared from our television and computer screens twenty-four hours a day. The *Wall Street Journal* article even mentioned my good friend, Dr. Donald Henderson.

> These were the sorts of calamities foreseen long before 2020 by eminent epidemiologists such as Donald Henderson, who directed the successful international effort to eradicate smallpox. In 2006 he and colleagues at the University of Pittsburgh considered an array of proposed measures to deal with a virus as deadly as the 1918 Spanish Flu.
>
> Should schools be closed? Should everyone wear face masks in public places? Should those exposed to an infection be required to quarantine at

home? Should public-health officials rely on computer models of viral spread to impose strict limitations on people's movements? In each case, the answer was no, because there was no evidence these measures would make a significant difference.[63]

Can you imagine how difficult it was for me, knowing the type of work I did, to be confronted daily with the madness that was the response of Drs. Fauci, Birx, and Collins to the pandemic? This was my area of expertise, and they didn't want to hear from me in the slightest. They wanted to hear from no one who would contradict their, what I would term, "lockdown lunacy." It was a ZERO-COVID mentality that guided all policy decisions flowing from Drs. Fauci, Birx, and Collins, despite the clear evidence that ZERO-COVID was impossible and would end up ruining economies and destroying lives. Importantly, you have no society to emerge to and you delay the inevitable, denying the population any movement toward natural immunity (baseline immunity) and population-level herd immunity.

How much did these agents of panic, misinformation, and terror hate it when we brought up the actual science? I wrote to the task force members routinely. I gave them the science. We gave them the science. They detested Dr. Atlas. They detested me. They detested anyone who presented the science and, in my opinion, were averse to the science. They were never following the science, from all I saw.

I will quote directly from the book published by Dr. Deborah Birx, *Silent Invasion*, of what she thought of me, especially of my alliance with Dr. Scott Atlas, of Stanford University.

That Scott Atlas had found a receptive listener in Paul Alexander at HHS was not surprising. Michael Caputo, who was appointed assistant secretary for public affairs in early spring 2020, had hired Alexander as his scientific advisor, poaching him from his position with the Infectious Disease Society of America in DC. Alexander had engaged in a campaign to wrest control of the public messaging from the scientists and public health officials. Centering many of his efforts on the CDC and the NIH, he tried to exert influence over Bob Redfield and Tony Fauci. Eventually, the House Select Subcommittee would see some of the emails Alexander wrote to both doctors, accusing them of fear mongering and stating that the CDC's Dr. Anne Schuchat was out to embarrass the president.

Alexander would leave his position in September 2020, but not before Assistant Secretary Caputo showed his support for various conspiracy theories (among them, that the CDC had created a resistance unit to oppose the president and his coronavirus positions) and not before he accused various scientists of sedition. As bad as the pandemic had gotten, Paul Alexander openly expressed his support for natural/herd immunity, stating of young people and those at lower risk, "We want them infected."

For his part, Alexander was a credentialed public health policy professional, but though he had a PhD and had graduated with a master's degree from Oxford, he was an unpaid, part-time professor at McMaster University. Scott Atlas shared the same set of beliefs, but had better credentials, affiliations with Stanford and the Hoover Institution. Whether Paul Alexander realized this or not, along with John Rader, Michael Caputo, and Fox News, he had managed to put Scott Atlas more prominently in front of the president and his senior advisors.[64]

I genuinely appreciate it when my adversaries are honest about a situation. There are some things I can quibble about, such as not noting that my doctorate from McMaster's University was in evidence-based medicine, and that McMaster's is the world's leader in that field. But for the most part what she claims about me is accurate. I was working with others like Scott Atlas to wrest control of the public messaging away from her and Fauci because they were ruining the country. Dr. Birx and Dr. Fauci misled and continued to mislead the president at every turn. In my opinion, Fauci and Birx were incompetent and inept on public health and epidemiology. As a more measured look is being taken at their actions during the pandemic, the mistakes on lockdowns, masking, and forced vaccinations become even clearer. I was vindicated; Scott Atlas is vindicated.

Michael Caputo did claim there was a resistance unit inside the CDC to oppose the president and his coronavirus positions, a claim which I also believe based on what I saw. (Isn't it curious that Dr. Birx simply notes the claim, but doesn't dispute it?)

And to give Dr. Birx credit, a few pages later she does reiterate her belief that herd immunity would have caused many more deaths than the way the pandemic was handled. However, what she fails to do is provide any evidence for her position, or to account for the new data that has subsequently been published, showing lockdowns were a mistake that inflicted untold suffering on the world.

If the evidence was so strong that lockdowns and masking were unlikely to be helpful, and perhaps we should have engaged in different strategies (like maybe telling people to lose a few pounds, exercise every day, lower stress, and maybe have some hydroxychloroquine, ivermectin, and zinc in their medicine cabinet), what caused Drs. Fauci, Birx, and Collins to veer so dramatically from "the science?"

Computers. (And, in my opinion, a determination to make Trump look bad at all costs, even if the cost was hundreds of thousands of human lives.)

Not data from the front lines of the epidemic. Not listening to doctors and nurses in the field, researchers cataloguing who was likely to have serious consequences from the infection, and those who recovered with little difficulty.

That is science.

That is medicine.

Instead, Drs. Fauci, Birx, and Collins took their marching orders from scientists who plugged a bunch of wild conjectures into a computer and had it spit out possible scenarios. In January and February 2020, genuine scientists had the plans which should have been followed. The *Wall Street Journal* editorial picked up the thread:

> But those plans were abruptly discarded in March 2020, when computer modelers in England announced that a lockdown like China's was the only way to avert doomsday. As Henderson had warned, the computer model's projections—such as 30 Covid patients for every available bed in intensive-care units—proved to be absurdly wrong. Just as the British planners had predicted, it was impossible to halt the virus. A few isolated places managed to keep the virus out with border closures and draconian lockdowns, but the virus quickly spread once they opened up. China's hopeless fantasy of "Zero Covid" became a humanitarian nightmare.[65]

We are told that this change of heart has come about because of two recent reviews, "one by Health Resources and Services Administration Official Jim Macrae into the CDC's pandemic response and another by CDC Chief of Staff Sherri Berger into agency operations."[66]

In other words, the agency responsible for this disaster investigated its own mismanagement. Does your boss let you do that? Do you think that's going to get us an unbiased look at these agencies?

In my opinion, the only things that are likely to be uncovered are the failings, which are undeniable. Here's how current CDC Director Dr. Rachel Walensky responded to the findings of the report. "For 75 years, CDC and public health have been preparing for COVID-19, and in our big moment, our performance did not reliably meet expectations," Walensky said in a statement. "As a longtime admirer of this agency and a champion for public health, I want us all to do better."[67] She was doubling down on their bad performance, claiming it was more of a messaging problem than ill intent, or perhaps just basic stupidity.

> There is consensus within the CDC that it "needs to make some changes for how it communicates and how it operates to be faster, to be nimbler, to use more plain-spoken language," said a CDC official, who was granted anonymity to discuss the changes before they were announced.
>
> "People work incredibly, incredibly hard and care deeply about trying to make sure that the American people have the right information," the official said. "Maybe the way that a lot of the [Covid-19] response was structured, and some of the incentives that people have here, are just not aligned properly to really put the focus toward getting information to people quickly and how that information can benefit Americans' health."[68]

Despite all the evidence that has recently accumulated about the failure of their measures, Dr. Walensky would go on to say that we should have all locked down harder.

However, the *Wall Street Journal* editorial wasn't having any of it and gave it to Dr. Fauci and Dr. Walensky with both barrels.

> Lockdowns and mask mandates were the most radical experiment in the history of public health, but Dr. Walensky isn't alone in thinking they failed because they didn't go far enough. Anthony Fauci, chief medical advisor to the president, recently said there should have been "much, much more stringent restrictions" early in the pandemic. The World Health Organization is revising its official guidance to call for stricter lockdown measures in the next pandemic, and it is even seeking a new treaty that would compel nations to adopt them. The World Economic Forum hails the Covid lockdowns as the model for a "Great Reset" empowering technocrats to dictate policies worldwide.[69]

Dr. Fauci and company must surely be the most out of touch with reality people on the entire planet. When their mistakes are pointed out, they complain and whine that they work "incredibly hard" and attribute it to their failure to use "plain-spoken language." It's as if we're all a bunch of Appalachian rednecks who, between all the science talk, would appreciate it if they burped and farted a few times, and maybe scratched their balls to show they're just like us regular folk.

Fortunately, the editorial writer at the *Wall Street Journal* is intelligent enough to translate our inarticulate grunts into something more elevated.

> It was bad enough that Dr. Fauci, the CDC and the WHO ignored the best scientific advice at the start of this pandemic. It's sociopathic for them to promote a worse catastrophe for future outbreaks. If a drug company behaved this way, ignoring evidence while marketing an ineffective treatment with fatal side effects, its executives would be facing lawsuits, bankruptcy, and probably criminal charges. Dr. Fauci and his fellow public officials can't easily be sued, but they need to be put out of business long before the next pandemic.[70]

From the time I was a young man, I have been in love with America. I have loved its founding principles, the optimism of its people, the example of freedom it has shown to the world, and its bedrock belief that the citizens are the genuine rulers of this land.

But I cannot say I love its government.

Aside from a few genuinely smart, brave, and patriotic individuals (whose tenure was often short like mine because they were driven out of their positions), for the most part the government is run by risk-averse people who glory in their own power and consider themselves the true rulers of this country. They also largely comprise what we would call the bureaucratic Deep State.

They have now released two reports which claim to tell the story of what went wrong with the COVID-19 response.

I will tell you what I observed, in the days, weeks, and months after I met with two senators of different parties and a congressman, who said they wanted me to use my evidence-based science skills to write a report on the CDC, which would, "take it down to the studs."

The attempt to force lockdowns on the healthy population, close the schools when children were not at risk, and engage in a reckless plan to rush an untested vaccine (gene therapy) on the public had just one goal; to prevent a second Trump administration. It was nothing less than treason against the American republic, led by the public health scientists and the Deep State.

CHAPTER THREE

Entering the Washington, DC, Quagmire

My first official day of work for the COVID-19 task force was around May 10, 2020.

By that time, the country had been shut down for nearly two months, with no end in sight. The medical and scientific communities might be forgiven for initially expressing support for a policy of "fifteen days to flatten the curve" as the response to the terrifying unknowns of SARS-CoV-2. If you had asked me my opinion when they told us that students and workers would be home for a little more than two weeks, I probably would have shrugged and said, "Not necessarily a bad idea. It doesn't have strong support in the literature, but we might be looking at an unprecedented situation. It is reasonable that this short period will allow us the time to get our arms around this and fully understand what we were dealing with." If we assume each day of lockdown requires ten days of normality to repair, as some economists have predicted, we were looking at 150 days to get back to where we were before the outbreak.

We were taking a risk with people's livelihoods and schooling, but it could be justified as a reasonable risk. The problem, though, was that we quickly had the data and evidence to show us very, very early on that COVID was amenable to risk stratification with a steep age-risk curve, where baseline risk predicted outcome. We quickly knew who was at risk and how to manage and treat. It was the hardening and prolonging of the lockdown policies that was catastrophic.

As I've mentioned before, my office was on the sixth floor of HHS, the seventh floor was the home of "Operation Warp Speed," and around the fifth floor below me was the special conference rooms where we often had our meetings.

My office door had a nameplate which read, "Dr. Paul Alexander—Senior Pandemic Adviser," I had a nice desk and cabinets and, directly in front of my desk, a television was on the wall. We were expected to always have it on to *CNN* or *Fox News* to keep abreast of breaking news. The presence of televisions in the offices of top officials at the various public health agencies was something to which I never grew accustomed. I'm wondering if some future presidential candidate might make as part of his or her platform, "No televisions in government offices so bureaucrats can see how they're coming across on the news!"

Michael Caputo's office was just down the hall, and he would be my main ally against the craziness of the HHS, the communication wing, and the COVID-19 task force. The alphabet agencies had sub-offices in HHS, and, as such, Dr. Redfield, Dr. Hahn, Dr. Giroir, etc. spent as much time in HHS as their other offices. One might say Caputo and I were the dynamic duo for Trump against the bad science that was being practiced by Fauci and Birx. Caputo did not have a scientific background but was highly intelligent, very technical, and always made sure he did his homework. I admired him for how prepared he was, and he knew about the science as if you were talking to a scientist. He has an insatiable mind.

We'd actually connected prior to my time in government. As a well-known conservative, he often guest-hosted for talk show host Sandy Beach (and I think Rush Limbaugh), and I was a regular caller. Rush liked me, as I was a scientist who would often talk about the distortions related to global warming, and, as a result, Caputo came to like me as well over the times I called into him, mainly on the *Sandy Beach Show* (WBEN New York).

Caputo also did something that I'm sure the Deep State knows all too well, but that my readers need to appreciate. Caputo oversaw messaging, especially from the National Institutes of Health, but realized he didn't have scientific training and was going to be interacting with both politicians and scientists. He had the politicians down cold, but not the scientists.

From the start, Caputo had me sit in on many of his important phone calls, whether it was with Anthony Fauci, Deborah Birx, Robert Redfield, the president's gatekeeper Kelly Anne Conway, Chief of Staff Mark Meadows, or even Trump himself. "I just want you to sit there and be quiet," Caputo told me. "Then when the call is over, I want you to tell me whether you agree with what's being said about the science." My role was to ensure that the science was clear, that he was not being misled, and to help support him and any of the bosses with the aim constantly of bringing the best decision-making to the table. Always to benefit the American people. While I would give him technical information and background about the science, he would do the same for me about the politicians. I enjoyed working for Michael for I also saw the kindness and humanity in him, beyond politics, especially his exceptionality as a husband and father. I was glad I decided to work with him in this important time in our history.

I sometimes picked Caputo up in the morning at his Washington, DC, townhome, which was located right next to that of Senator Majority Leader Mitch McConnell, whom I often saw early in the morning and would say hello. Caputo also loved to exercise and would ride or walk to the office. I spent the majority of my days with Caputo, we often had lunch together, and I would gladly work with him again. I think Caputo is one of the best people I have ever met in terms of character, love of nation, decency, and work ethic. He was also in the military. You have to know him and work with him to understand that he is very loyal and places high value on honor and moral code. As do I.

I was being given an unprecedented opportunity to watch our government in action, along with running commentary by somebody who understood its soul. I learned a lot from Caputo.

* * *

Before I begin to tell you about the personalities of those with whom I interacted, I must tell you about the circumstances of my employment. Imagine you are me, originally from the Caribbean islands. You're summoned to Washington, DC, told in a secret meeting with two United States senators and a congressman soon after arriving that your key role while at HHS has more to do with the examination and remaking of the CDC to help expose its corruption by focusing on its published guidance and reports, and told in a roundabout manner that it may or may not be a request from President Trump and that he may or may not be knowledgeable about it (for sure folk in the Oval Office were, but I was not told who). Add to that the fact you left a decent paying job at the World Health Organization and the Pan American Health Organization (PAHO). While you consider this a position of public service that you are happy to take, you figure you're going to get paid a certain amount of money to help you with your bills, right? I felt that I would be moving to Washington, DC, and receiving the support and help of the government bureaucracy to settle in and to do my job. I think I'm a pretty decent looking fellow and all, with brown/dark skin, a foreign accent, and an impressive academic pedigree, but that wasn't going to pay the rent or buy groceries for my family.

The harassment began with not wanting to complete my human resources paperwork or paying my salary.

When this became clear to me after about two weeks on the job, I contacted Human Resources at HHS and asked the reason for the delay.

They told me my paperwork had not been processed. This went on for approximately four months.

Finally, it got solved through the intervention of some military guys working on the seventh floor above me in "Operation Warp Speed" (OWS). One of them approached me after a meeting and basically said, "We understand you're not getting paid." At this time, I was participating in OWS meetings when I was invited, and, in those meetings, there was always a heavy military presence.

"That's right," I replied. "Something about my paperwork not being processed."

"I think we can get you paid from some of our funds," he said.

"Can't you get me my backpay as well?" I asked.

"Probably not," he answered.

In my last six weeks working for HHS in support of the COVID-19 task force, I did get my promised salary. But I didn't get my backpay. The Human Resources people I spoke to also said there would be no backpay.

They said it was because I was erroneously listed as a "volunteer."

But I don't believe that was true.

It was just another way to denigrate my contribution and my credentials. Experts are generally paid significant amounts of money for their expertise. I was such an expert in evidence-based medicine.

To me this was the first piece of evidence that the bureaucratic Deep State knew exactly what my core role in Washington was as it pertained to the CDC task. It was likely that information was being leaked. This was not surprising given Washington operates only on leaks and the timeliness of leaks.

* * *

As mentioned, my office was on the sixth floor of HHS, just a few steps away from the office of Alex Azar, the head of HHS under Trump.

In Washington, DC, they always say that proximity equals power, but even more important than being close to Azar's office was the fact that the home of Operation Warp Speed was just above me. Operation Warp Speed was the Trump plan to bring out a COVID-19 vaccine in record time. The effort was staffed by scientists from various pharmaceutical companies, as well as from government labs. The OWS effort was also mostly comprised of military, and I mean that soldiers in full military uniform occupied the seventh floor. Note that the military had a strong role in the logistics aspect of OWS. This is from the HHS website explaining the program:

> OWS is a partnership among components of the Department of Health and Human Services (HHS), including the Centers for Disease Control and Prevention (CDC), the Food and Drug Administration (FDA), the National Institutes of Health (NIH), and the Biomedical Advanced Research and Development Authority (BARDA), and the Department of Defense (DoD). OWS engages with private firms and other federal agencies, including the Department of Agriculture, the Department of Energy, and the Department of Veteran Affairs. It will coordinate existing HHS-wide efforts, including the NIH's Accelerating COVID-19 Therapeutic Interventions and Vaccines (ACTIV) partnership, NIH's Rapid Acceleration of Diagnostics (RADx) initiative, and work by BARDA.[71]

This was a massive operation on the part of the federal government, with many equating it to the Manhattan Project of World War II to build an atomic bomb before the Nazis, although it was more bad luck for the Japanese.

One of the things my Caribbean accent has always done is attract people to me. It sounds pleasant to their ears. which may have made some of my critical comments about the COVID-19 response that I made in meetings and in teleconferences go down a little easier with some people, and not with others.

You have to understand scientists are trained to be collaborative and collegial, discussing ideas in an open forum, and not letting it get personal. And when I started, that was pretty much the feeling of the place, even when we had differing opinions. I sat in on many meetings of Operation Warp Speed, and freely expressed my opinions about what they were doing. I'm a friendly person, and when you're working eighteen to twenty hours a day in what seems like war-time conditions, you tend to form strong bonds with people.

I'd often get coffee with good scientists from the NIH, CDC, FDA, Pfizer, and Moderna, and when they'd see we had some privacy, they would lean across the table and whisper in a low voice, "Paul, we know what you're saying, and we agree with you. But if we said it, we'd lose our lives and careers. But don't you stop. We want you to tell the truth. Keep standing up. We want to speak out but we cannot out of fear of retaliation and loss of income."

The quandary in which these scientists found themselves was a genuine one. The media had whipped the COVID-19 fear into a red-hot panic, and, when people are terrified, they can't think straight. They simply want to feel safe again. Not many people appreciate the extent to which Operation Warp Speed was a combined civilian and military operation. Again, this is from HHS's own website explaining the organizational structure of Operation Warp Speed:

> HHS Secretary Alex Azar and Defense Secretary Mark Esper oversee OWS, with Dr. Moncef Slaoui designated as chief advisor and General Gustave F. Perna confirmed as the chief operating officer. To allow these OWS leaders to focus on operational work, in the near future the program will be announcing separate points of contact, with deep expertise and involvement in the program, for communicating with Congress and the public.[72]

Let me go through some of these people, so you can understand that it's something of a misnomer to call it a civilian-military operation.

It's more accurate to describe it as a pharma-military operation.

I need to begin by saying that I don't know why President Trump picked Alex Azar to be the head of HHS. He doesn't have a scientific background. He is an attorney, lobbyist, and pharmaceutical executive. Prior to joining the Trump Administration, he was the president of the US division of Eli Lilly and Company, a pharmaceutical drug company. In my opinion, it's like putting convicted fraudster Bernie Madoff in charge of the Treasury Department. The only thing a guy like that will do is make sure the interests of the pharmaceutical industry are well-served.

In my interactions with Azar, he always seemed like a smooth operator. I met several times and sat down with him to have casual conversations, and I just never left those meetings with a good feeling about the man. He always had an entourage around him and always seemed attuned to the political winds.

Then you had Secretary of Defense Mike Esper heading up the effort. Who is Mike Esper? In addition to his time in the military (1986–2007), he lobbied the government for the defense contractor, Raytheon. Let's just say it.

His job was to get government grants to benefit the military industrial complex.

This is not a public servant. He's an agent of the defense contractors, and President Trump gave him the opportunity to jump sides and shovel money to his good friends in the defense industries.

General Gustave Perna, chief operating officer of OWS, was a long-time military man, specializing in logistics for the Army, mainly concerned with logistics. That means he was exactly the type of person that Michael Esper, in his previous position as a lobbyist for Raytheon, would love to get all excited about a new weapons system.

Then you had people like Dr. Moncef Slaoui, one of the four people responsible for Operation Warp Speed. Who is Slaoui? Slaoui is the guy who headed vaccine development for Moderna. I went to Slaoui's office several times, sat down with him one-on-one, and shared the science that I thought he should know. I also sent him breaking updated science that I thought he should be aware of, all generally pertaining to the safety aspects of the vaccines.

I have to make one more comment about the staff at Operation Warp Speed. I think I have a pretty good sense of those who were scientists and those who were either pharma executives or military men. But there was another group there, and they didn't talk much. If you did talk to them or ask what they did, you'd get something official sounding, which didn't make much sense.

I strongly believe they were intelligence people—spooks, in other words—and I couldn't imagine why they were so interested in the development of a medical product. The only reason that made sense to me is they wanted to monitor what we might be saying to each other about the origin of the virus, whether it was a Chinese escape or planned release, as well as questioning what role the United States might have played in the development of such a pathogen. If that was the intention, it certainly worked, because I never heard any member of Operation Warp Speed asking any questions about the origins of SARS-CoV-2. But do not get me wrong: I am a great admirer of the military and people who serve our fine nation, yet I was always intrigued by the makeup of the people on the seventh floor of HHS, directly inside of the OWS offices.

I greatly admired some of the people on the task force, such as Dr. Brett Giroir, who oversaw COVID-19 testing. He was handed a tremendous headache because the first COVID test developed by the CDC was an absolute disaster. They went to the labs to validate the test, and it failed. That's why early in the crisis the tests were stopped for five weeks until an investigation was completed and a more accurate test could be developed. Maybe such a foul-up was inevitable, but that five-week delay because of the CDC's failed COVID test genuinely harmed our country as it allowed the virus to be seeded across the eastern and western parts of the United States unchecked. Americans to this day do not understand how much the CDC botched the initial testing rollout and how much it hurt the response.

I tried to develop a good relationship with Dr. Stephen Hahn, who was head of the FDA. I was in many meetings with him, one-on-one, as well as in larger groups. He always seemed to welcome the information I gave him, as well as my views and opinions on the COVID response (especially the copious studies and science I shared with him and the FDA as they were published, the aim being to ensure he was up to date on the science), and because of my direct contacts in the research community, I could inform him in advance of upcoming publications. I had a good deal of respect for him, and he seemed to be a fine family man, but seemed hesitant to challenge the consensus, even when there was strong evidence it was wrong. I did find he was weak as the leader of the FDA, and I was genuinely disappointed when he took an executive position with Moderna after leaving the task force. Though I have nothing ill to say about him—especially about his character—it made me question his true loyalties. In my opinion, Dr. Hahn lacked the assertiveness to stand up to the Deep State of the FDA (and HHS). Maybe he was just playing a game with me, figuring out how the research findings might eventually cut into Big Pharma profits.

* * *

Probably the most surprising relationship I had during my time with the COVID-19 task force was with Dr. Robert Redfield, Trump's pick to be head of the CDC.

Dr. Redfield was a man of faith, a good family man, and talked a lot about his children, of whom he was enormously proud.

I was told by others that when Dr. Redfield was appointed, there was a good deal of anger at the agency because he wasn't in their club. My understanding was he did not agree with a lot of the policies and decisions coming from the highest levels of the CDC. In other words: from inside his shop. Eventually our relationship deepened to the point where pretty much every day I could expect Dr. Redfield to drop by our office and talk about all the crap he was dealing with in terms of CDC politics. And there was a lot.

As an example of the topics we might talk about, he once told me he was very concerned with the quality, accuracy, and completeness of the reporting the CDC was doing on the pandemic. He explained that the main editor of the CDC reports was very capable and doing a fantastic job, but the scientists who were giving her the reports were doing so with incomplete data, and a definite political slant. I always felt he was trying hard to improve the CDC and to ensure its guidance was complete, fair, balanced, forthcoming, and accurate. He strived for high-quality work and was not getting it from the CDC.

In several ways, through the three politicians who were contacting me for updates on my CDC report, it was intimated to me that Dr. Redfield knew at least some of our concerns and was an ally. I never directly discussed the matter with him, but he seemed as frustrated as all of us were by the actions of the CDC. Dr. Redfield, in my opinion, seemed trapped between how he wanted the CDC to operate optimally for the American people and how the Deep State bureaucracy and technocracy was running it. To me, it was almost as if there were two CDCs: the one Dr. Redfield thought he was leading, and the one being led by technocratic bureaucrats.

I remember after leaving a meeting with Dr. Redfield and some of my bosses that Dr. Redfield and I had a discussion walking in the hallway about the six-foot social distancing rules. I asked if he could explain to me and show me the science they relied on for publishing the six-foot recommendation, which was then becoming ubiquitous across the country, with stickers on the ground at ATM machines, grocery checkouts, and every place you could possibly imagine.

He fixed me with a stare, chuckled, and then said, "What science, Paul? There is no science. Other nations say one foot, some say nine, others twelve. We felt six would sound better, so we went with it. We made it up."

That was the COVID-19 fantasyland we were all living in, seeing the lies, but unable to stop them.

Although Dr. Redfield was a good man, he seemed clearly outmaneuvered by the Washington, DC, power players. In my opinion, I would work with him anytime, but he was just too weak to handle the Deep State at the CDC.

It was a dangerous game we were all playing, and our opponents had much more experience at cutthroat politics.

* * *

I entered the government as a GS-15 government employee, which is the pay grade generally reserved for top-level positions such as supervisors, experienced technical specialists, and top professionals with advanced degrees. Above the standard GS-15 is senior level service, which is given to the governments' most renowned researchers.

What my GS-15 clearance also gave me was access to the high-level lunchroom on the sixth floor, reserved for members of congress, the senate, or top officials of NIH, CDC, FDA, or other branches of the government.

It was in that lunchroom that I received probably the greatest education about the ways of Washington, DC, during the Trump administration. It became very clear to me that many persons in the government, at the highest levels and on both sides, were constantly working against President Trump. The NIH numbers a little more than eighteen thousand employees and had been staffed by the Bush and Obama administrations, both of whom despised Donald Trump. Maybe there were a couple hundred people in the entire building who supported Trump, and just a few among top management. I would say five.

As much as you will hear liberals decry racism, they are among its most dedicated adherents. I'm an olive-skinned man of Middle Eastern descent and speak with a thick Caribbean accent, so it was common for people who'd meet me in that lunchroom to say, "I heard you on the COVID conference calls and figured you were a brother! I didn't expect you to be white!"

What could I do but shrug my shoulders and say, "Sorry, that's me."

However, a funny thing usually happened after such comments. Even though they knew I'd been hired by President Trump's team, they'd just look at me and listen to my voice, and it simply didn't compute that I was conservative.

Then they'd start talking to me like I was one of them, or maybe I was somebody new to the club. I might not be sophisticated in the ways of the Capitol, in the world of Washington, DC, and its politics, but they would certainly aid me in my education. Surely, this charming, misguided foreigner could eventually be shown the error of his ways.

"None of us in the bureaucracy wanted Trump," they'd say, sweeping a hand across the room. That's how they referred to themselves. Not as the "Deep State," but as "the bureaucracy," which sounds about as terrifying as anything George Orwell might have written. The conversation would normally continue in this vein, and it was clear people were quite comfortable with me. What was unsettling to me is that there were people I knew who were actually appointed by the Trump team as political appointees or who worked as federal employees yet seemed very averse to President Trump. I could not understand why they would take a position yet not be loyal and supportive.

It shocked me when I was told, "And we in the bureaucracy are dedicated to making every day of his life a living hell. When Americans watch the evening news, all they'll see will be another day of the country not working under this president. Ungovernable, unmanageable, chaotic, infections going up and up. Americans will want anyone but Trump and we are doing it, for we have all the health agencies like CDC and NIH and FDA working with us. We have Fauci with us, we have Birx with us. How could he win? Our job is to make the pandemic response appear to be a disaster, and we coordinate roughly every day across the different agencies to make it look that way and achieve the goal." They also were fond of using the expression, "we in the bureaucracy," which sounded to me like a perversion of the phrase "We the people" that opens the United States Constitution.

I'd play it close to the vest, not agreeing or disagreeing with them and sometimes jabbing a little with them, but they seemed to take my comments as something cute, like those of a precocious ten-year-old.

For as much as President Trump talked about the bureaucracy, I don't think he ever understood the depth of their hatred toward him. For that matter, I don't think most of the citizens understand it, either. They believe the comforting lie that when the election is over, the politicians and bureaucrats simply get back to doing the people's business.

That is the farthest thing from the truth.

They are partisans and rarely think of doing what's best for the citizens.

* * *

I did not begin my time at HHS in support of the COVID-19 task force with a high opinion of Dr. Anthony Fauci.

I had extensively reviewed his performance during the HIV/AIDS epidemic of the 1980s and found many deficiencies in his performance. I held him responsible for many thousands of unnecessary deaths of gay men in the United States and around the world because of his hesitation in promoting a decades-long approved medication, Bactrim. This drug prevented the development of the pneumocystis pneumonia (PCP), which often precipitated the death of HIV/AIDS patients. Dr. Fauci, however, promoted AZT, a drug that was highly toxic at the levels in which it was prescribed at the time.

Does that sound like a familiar pattern we have observed Dr. Fauci to follow in COVID-19? Suppress the use of a long-approved medication, and in its place push a very deadly one? A 2009 publication by one of the doctors, Dr. Joseph Sonnabend, who treated some of the first HIV/AIDS patients in New York City, recalls these twin mistakes by Dr. Fauci. As he recalled:

> It's a sad, largely neglected history. Many lives were shortened before 1989 when interventions to prevent this type of pneumonia were finally recommended by government officials for people with AIDS.
>
> Thankfully, with the widespread use of antiretroviral drugs, many people may not be too familiar with this opportunistic infection. It most certainly has not gone away, but in the 1980s pneumocystis pneumonia was what most commonly killed people with AIDS.[73]

The gay population endured years of terror and hopelessness when that burden could have been greatly lifted by the simple use of Bactrim for those at risk. Remember that HIV/AIDS at its height never affected more than 1 percent of the population. This was a crisis that Dr. Fauci should have been able to manage with relative ease. How many deaths did this delay cost? Dr. Sonnabend provides an answer:

> To get an extent of the fatalities this pneumonia caused, Michael Callen [an early AIDS activist] asked a CDC statistician in 1989 how many AIDS related PCP [pneumocystis pneumonia] deaths had occurred since the beginning of the epidemic. As of February 20th, 1989, 30,354 Americans had died of AIDS-associated PCP. The year is significant as it was then that the CDC finally issued recommendations for the prevention of PCP, using a drug that had been known to prevent this kind of pneumonia since 1977.

The drug in question is Bactrim, also known as Septra, Septrin, or co-trimoxazole. It is actually a combination of two drugs, trimethoprim and sulfamethoxysole. It has been available as an inexpensive generic product for many years.[74]

Is this approach reminiscent of how Fauci acted during the COVID-19 crisis? Hydroxychloroquine (HCQ), a drug useful for treating COVID, had been available for decades and was listed by WHO as an "essential medicine." This issue was close to Dr. Sonnabend's heart as Michael Callen was his patient and had even traveled to Washington, DC, to meet with Dr. Fauci.

Michael, with other activists, met with Dr. Fauci in May of 1987. Michael was insistent in asking for recommendations to prevent PCP in people with AIDS. Michael wrote the following in relation to this meeting:

"It is particularly galling to me that 16,929 of the 30,534 unnecessary PCP deaths occurred since May of 1987, the date on which I and other activists met with Dr. Anthony Fauci (the closest person we have to an AIDS czar) to ask him—no, to *beg* him—to issue interim guidelines urging patients to prophylax those patients deemed at high risk for PCP. He steadfastly refused to issue such guidelines. His reason? No data. As a result, many more people died of PCP who didn't have to."[75]

Why is it that Dr. Fauci failed to issue even temporary guidelines for a medication that had already been deemed safe and was being used for pneumocystis pneumonia in other groups of high-risk individuals? At the time, many AIDS activists were claiming members of the government didn't care about the gay population or were actively slow-walking possible treatments to dilute the growing political power of the gay movement. At the time, these sounded like wild claims, but with the passage of time I think it would probably be a good time to revisit such allegations and determine if they have any merit.

Sonnabend summarized the Bactrim situation:

In the early days of the epidemic, we could not know which patients were at risk for PCP. We had to learn that [the level of] 200 CD4 cells was the dangerous threshold, below which there was a substantial risk of infection. But well before this we were perfectly able to target a population at great risk for PCP: these were people who had experienced one attack already. They were almost certainly going to experience another one but their protection was not considered a matter of urgency by the federal AIDS medical leadership. Of course, in the absence of effective treatments for HIV disease, preventing PCP would have been a life extending rather than a lifesaving intervention.

Another curious and indefensible objection to PCP prophylaxis was raised by Dr. Samuel Broder, who was then head of the National Cancer Institute. He felt it justifiable to *discourage* the use of PCP prophylaxis on the grounds that the introduction of AZT would make the practice redundant! This object was raised in the complete absence of any evidence that AZT could prevent PCP in a significant and durable fashion, if at all.[76]

I genuinely wish that every American would take a deep dive into the federal government's response to the HIV/AIDS crisis. In my opinion, what was done to the gay community by our public health authorities, led by Dr. Fauci, was an absolute crime, dwarfed only by their response during the COVID-19 crisis.

The analogy with Bactrim and HCQ is that Dr. Fauci was hesitant to promote Bactrim in lieu of AZT, which was toxic and killed thousands of gay males. Similarly, Dr. Fauci was averse to HCQ, which was an approved therapeutic drug for safe use in lupus and malarial cases. Instead, he promoted the use of remdesivir, which turned out to be liver and kidney toxic with no effective antiviral effect.

Specifically, while Dr. Fauci was downplaying the effectiveness of Bactrim, he was stoking AIDS fears and promoting the unsafe use of AZT, which was actively killing AIDS patients. Robert F. Kennedy Jr. profiled Dr. Fauci in his excellent book, *The Real Anthony Fauci: Bill Gates, Big Pharma, and the Global War on Health and Democracy*, and had this to say about Fauci's work during the HIV/AIDS epidemic.

> Despite pleas from patients, their doctors, and advocates, despite the vast financial windfalls flowing to his agency from the HIV community's adept lobbying, Dr. Fauci refused to meet with the AIDS community leadership during his first three years as America's "AIDS czar." That reticence further soured Dr. Fauci's already difficult relationships with the community he was responsible to serve.
>
> It was a hardwired reflex at NIAID [National Institute of Allergy and Infectious Diseases] to exaggerate public fear of pandemics, and Dr. Fauci's first instinct as national AIDS czar had been to stoke contagion terror. He had made himself a villain among AIDS activists with a fear-mongering 1983 article in the *Journal of the American Medical Association* warning that AIDS

could be spread by casual contact. At the time, AIDS was almost exclusive to intravenous drug users and males who had sex with other males, but Dr. Fauci incorrectly warned of the "possibility that routine, close contact, as within a family household, can spread the disease." Given that "nonsexual, non-blood borne transmission is possible," Fauci wrote, "the scope of the syndrome may be enormous." In his history of the AIDS crisis, *And the Band Played On*, author Randy Shilts reports that the world's leading AIDS expert, Arye Rubenstein, was "astounded" at Fauci's "stupidity" because his statement did not reflect the contemporary scientific knowledge. The best scientific evidence suggested the infectivity of HIV, even in intimate contact, to be so negligible as to be incapable of sustaining a general epidemic.[77]

They say that criminals start with small crimes. Serial killers don't start with people, but start by torturing and killing small animals, then start moving up the chain until the system of hunting, inflicting pain and torture, the orgasmic release of the kill, and figuring out how to dispose of the evidence of the crime has been perfected, and they're ready to hunt the ultimate prey.

Are people terrified of epidemics?

You bet they are.

You don't see the virus, but it lives in your imagination. We are very concerned and afraid of contagion. You picture it landing on the surface of your cell like some spindly spacecraft, deploying its drill, which deposits its genetic payload into the nucleus of your cell, where it hijacks the cell into producing copies of the virus until it bursts from the cell like the monster of the *Alien* movies, exploding from the midsection of a hapless astronaut.

Science is supposed to reduce the fear.

The problem is taken from the realm of the mysterious into a series of manageable steps.

That's how I approach science. The key is I enjoy and seek out ways to share information so that all around me know what I know.

I want you to take responsibility for what should be done. Once I've shared my knowledge with you, you know as much as I do. Your decisions are likely to be just as informed and optimal as the ones I make.

I hope this explains why I consider Dr. Anthony Fauci to be such a destructive force in our world and oppose everything for which he stands. I do not dislike him, for he may be a great person. Yet it is his decision-making in this COVID emergency that has been catastrophically flawed and has been flat-out entirely wrong. His decisions have cost thousands of lives, from denial of early treatment, to the lockdown lunacy, as I would put it, to the vaccine and the mandates. In each of these cases, countless Americans and people around the globe died needlessly.

Consider the terror Dr. Fauci brought down on 1983 America with his irresponsible publication. The "science" said that the disease was one that afflicted intravenous drug users and men having sex with men (MSM), which often involves heavy skin-to-skin contact and intimacy, the micro tearing of intimate rectal tissue, especially in the anus, and thus the ready transfer of infected bodily fluids.

The virus itself was relatively fragile, and, if proper precautions and modifications had been taken (including by the leaders in the gay community at the time, who delayed their call to close the gay bath houses in urban areas), the problem would have been greatly reduced.

But Dr. Fauci's malfeasance wasn't limited to preventing life-extending medications from being used (i.e., Bactrim) or unnecessarily terrifying the public, but actively promoting the harmful and extensively toxic AIDS medication, AZT. Dr. Fauci concealed the terrible side-effects of the AZT drug (e.g., anemia), to benefit his collaboration with the Burrough/ Wellcome research team, a pharmaceutical entity. As Kennedy revealed in his Fauci book:

> Of those who got AZT, all suffered from its unspeakable toxicity. "A number of them . . . would have definitely died from anemia,' had the PIs [principal investigators] not given them blood transfusions to keep them alive, says Lauritsen. AZT causes anemia in every animal species studied, including human beings. In his book, *Poisoned by Prescription*, Lauritsen explains how "[p]atients taking AZT became anemic, and suffered low white blood cell counts accompanied by vomiting." FDA documents showed that everyone in the AZT group suffered severe toxicities and anemia, yet NIAID's official report listed no adverse effects among AZT recipients . . .
>
> Many of the patients would have died from the toxicities of AZT if they had not been given emergency blood transfusions," reports Lauritsen. "This is a serious adverse effect. That means literally they would have died from the poison. And yet the case report forms that showed up eventually would report no adverse effects. I mean, this is a type of dishonesty. It's hard to go any further than that."
>
> Dr. Wilner, who died in 1995, accused Dr. Fauci of using transfusions and other artifices to systematically conceal AZT's horrendous toxicity. "What do we have to say about the National Institutes of Health, when a private independent laboratory, found AZT to be 1,000 times more toxic than

the laboratory at NIH? We can understand a 5 percent error rate in a laboratory, even a ten percent error, but a 10,000 percent error or a 100,000 percent
error? That's fraud."[78]

How is it that Dr. Fauci and those who assisted him in these actions escaped
some form of legal prosecution? Even if we accept the proposition that
nobody gets fired from government work, one might figure a way to ease
him out of his position, as they did with Dr. Fauci's mentor, Dr. Robert
Gallo, when it was shown he had tried to steal credit for the isolation of the
HIV virus from French researcher Dr. Luc Montagnier.

But somehow Dr. Fauci managed to simply stay in his position past the
normal retirement age of sixty-five, blew through seventy, then seventy-five,
then somehow at the age of seventy-eight found himself in charge of the
COVID-19 task force.

I couldn't understand why Dr. Fauci didn't provide some simple messages to the various communities around the United States, which might
have drastically cut deaths. For example, the African Americans in the
United States died at much higher rates from COVID-19 than their white
counterparts. However, Africans in Africa had some of the lowest rates of
infection and death in the world. What accounted for the difference?

Well, Africa does have an extremely young population, which is something in their favor because the young are also at a much lower risk, but
also their dark skin was perfectly designed to get the proper amount of
vitamin D from strong rays of the equatorial sun. The United States has a
more northern, temperate climate, so if you've got dark skin, it's important
you get enough vitamin D from the sun, or more likely, should take some
vitamin D3 supplements. I know there are some critics who will accurately
say that a strong correlation has been established, but not causation.

Technically, that's true.

However, I put it in the same category as those of past centuries who
might have said, well, we've only established a correlation between vitamin
C and the development of scurvy in sailors on long ocean voyages, so we
shouldn't stock our ships with fruits like oranges and limes, which have
large amounts of vitamin C. It doesn't cost much to take a vitamin D sup-
plement, and even Dr. Fauci does it (even though he never made it an official
recommendation for the country). Here's a CNBC article from September
2020 about Dr. Fauci's own daily vitamin regimen, which includes vitamins
D and C.

According to Dr. Anthony Fauci, most "so-called immune-boosting supplements" actually do "nothing." However, there are two vitamins Fauci does recommend to help keep your immune system healthy.

"If you are deficient in Vitamin D, that does have an impact on your susceptibility to infection. So I would not mind recommending, and I do it myself taking vitamin D supplements," Fauci, 79, said during an Instagram Live on Thursday, when actress Jennifer Garner asked Fauci about immune-boosting supplements.

(In fact, researchers at the University of Chicago Medicine recently found a link between vitamin D deficiency and the likelihood of being infected with Covid-19—those with an untreated deficiency were more likely to test positive. "Vitamin D is important to the function of the immune system and vitamin D supplements have previously been shown to lower the risk of viral respiratory tract infections," David Meltzer, chief of hospital medicine at UChicago Medicine and lead author of the study said in a press release on Sept. 8.)[79]

Let's discuss first the absolute insanity of Dr. Fauci revealing this information in an Instagram Live interview with actress Jennifer Garner. Is Jennifer Garner a professional journalist, ready to pounce on lies and half-truths being peddled by the people she interviews?

I think not.

Does she have a medical degree, is she an academic scientist, or does she have any experience that might be relevant to this discussion?

Again, she does not.

I'm sure she's a lovely person, but the only real ass-kicking she's ever seen was when she played a secret agent on the long-running television series *Alias*. And I'm pretty certain none of that was real. (Just like I know Tom Cruise is not an actual Navy fighter pilot.)

And shouldn't Dr. Fauci have been saying this daily from the podium of the COVID-19 briefings? *Hey, citizens of the United States and people around the world, make sure you get your vitamin C and D levels tested and be sure to take your daily supplements just like I do?* But we didn't get anything like that. Yet with actress Jennifer Garner, we also get this important health advice from Dr. Fauci:

In addition to vitamin D, Fauci said that vitamin C is "a good antioxidant."
"So, if people want to take a gram or two at the most [of] vitamin C, that would be fine," he said.

(Vitamin C "contributes to immune defense by supporting various cellular functions" of the body's immune systems, according to a 2107 study published by The National Institutes of Health, and vitamin C also appears to prevent and treat "respiratory and systematic infections," according to researchers.)[80]

Did we get any of this information from those daily COVID-19 briefings? Check your vitamin C and D levels, America, maybe lose a few pounds, exercise more, and try to keep your stress low because high levels will affect the proper functioning of your immune system. Oh, and you should also probably know that this virus escaped from a lab in China that I helped to fund. It was incredibly important that the COVID-19 task force members use the podium to inform the public, yet they squandered the opportunity across two and a half years. Both the surgeon general for President Trump, and certainly the surgeon general for President Biden, have squandered it. They have never ever stood and told vulnerable parts of the population who needed the guidance and public service information anything that was of benefit. Just doom and gloom about the flawed epidemic curves based on the flawed over-sensitive, over-cycled RT-PCR test, and stronger, harder, longer lockdowns and school closures. Nothing of benefit.

I want to share with you some comments made by Senator Ron Johnson from the floor of the United States Senate on December 8, 2021, regarding how Dr. Fauci's actions may have led to more than half a million unnecessary deaths in the United States, and millions more around the world:

There are multiple medical experts who have looked at this, who are treating COVID, who are doing the research, who say upwards of 500,000 lives were needlessly lost because we ignored, and I would argue, sabotaged early treatment with cheap, generic, repurposed drugs.

In fact, the FDA completely trash-talked one of these repurposed drugs, a Nobel-Prize winning drug termed by the World Health Organization as a miracle drug, Ivermectin, calling it horse paste; saying: Come on, you all; you are not cows.

Fake news stories saying that people are lining up, clogging hospitals because of overdoses of Ivermectin, only to find out that is a completely false news story—like fake studies published in medical journals that had to be withdrawn 2 weeks later in the pandemic, also poisoned the use of some of these repurposed drugs.

Let's take a look at some facts, that when I go on media and describe the facts, I am censored by the COVID gods; I am removed from YouTube, as is sometimes the radio talk show host.

But let's look at the facts of the drugs versus the vaccine. Now, many of you will be shocked by this because this is all being censored. This information, this is not allowed. Again, there is one narrative; it is the narrative of the COVID gods. No second or third opinions are allowed. No questions are allowed to be asked, much less answered.

So let's take a look at Ivermectin. I have got two columns here: Total adverse events reported to either the FAERS system—the adverse events reporting system from the FDA for drugs – and the VAERS system—the vaccine adverse event system reported through the CDC.

So the top three. First of all, Ivermectin. Over 26 years—26 years of reporting—Ivermectin has about 3,756 adverse events reported through FAERS. So that is adverse events. In terms of death, it averages about 15 reports of death a year.

Now, let's get something straight here. There are two main criticisms of FAERS and VAERS. It doesn't prove causation. I get that. But is also dramatically understates the adverse events.

So, again, we are going to use this as a comparison.

Ivermectin: 15 deaths per year, on average, over 26 years of usage.

Hydroxychloroquine: 23,355 total adverse events over 26 years. On average, about 69 death reports per year.

How about the seasonal flu vaccine?

Again, 26 years' worth of history: 198,776 adverse events reported on VAERS, but an average of 80 deaths per year for the seasonal flu vaccine.

I look at these and I go: These are pretty safe drugs. No drug is 100 percent safe. No human body is exactly the same. But you have to look at these drugs as having a very safe and reliable safety profile.

So, if you have COVID—and let's face it, the current NIH guideline on COVID is to do nothing: go home, face it alone, isolate yourself, hope you don't get so sick you have to check yourself into a hospital

The only thing they are recommending for use is monoclonal antibodies. Try and get those. I have talked to so many constituents that haven't been able to. Either they are not sick enough or they have become too sick or it has taken too long

So, virtually, the NIH guideline continues to this day: Do nothing.

Now, a quick aside: How many other diseases is that the recommendation? Isn't it always better detection allows for early treatment, produces better outcomes?

Of course, that is what we recommend for every other disease, except COVID because Fauci ignored therapies and pushed vaccines. He has just been—he has got his blinders on. It is vaccines, vaccines, vaccines.

And then they scaremonger both Ivermectin and hydroxychloroquine.

I don't know. Are you afraid of those? If you have got COVID, would you give those a shot?

I certainly would. And, by the way, I am not a doctor; I am not a medical researcher. But I have been in contact with doctors who have the courage and compassion to treat. And so, when I have a friend or a constituent who calls me and says, "What can I do?" I refer them to a doctor who treats them.

And I have example after example of these things working, keeping people out of hospitals and certainly preventing death. I know it is anecdotal, but the evidence is mounting, and it is getting to the point of being irrefutable.

So now let's compare this to the drugs of choice of Dr. Fauci and the COVID gods. Let's take a look at Remdesivir. The studies were weak. They changed the endpoint of reducing death—because it didn't—to reducing days in the hospital. But they still rushed through the emergency use authorization, and it has been the treatment, because it is blessed by their COVID gods, that hospitals will apply.

Now, in fairness, hospitals also do dexamethasone. They will do other things—corticosteroids. Pierre Kory testified before my committee in May of 2020 about corticosteroids.

But Remdesevir is the big one; over $3,000 a dose when these cost 20 to 50 bucks, as part of a multidrug, multivitamin—vitamin D, zinc. Remdesivir: 6500 adverse events. 1,612 deaths so far since it got emergency use. That is an average of 921 a year. That is Remdesivir.

Now, let's look at the COVID vaccine, and this will shock you. It should shock you because nobody is talking about it. And when a guy like me talks about it, I get censored, I get vilified, I get attacked.

927,740 total adverse events. And remember, one of the criticisms is VAERS drastically understates the number of adverse events.

Total deaths: 19,532. Now again, I realize VAERS doesn't prove causation, but almost 6,000 of these worldwide deaths occurred on days 0, 1, or 2 following vaccination.

I know Fauci, I know Janet Woodcock [head of FDA], I know Francis Collins are not concerned about this. Other people who have been able to avoid the censors and see this, they are concerned about it. They are making those tough choices. They also realize COVID can be a deadly disease. They have to make an informed decision whether or not to get vaccinated.

Shouldn't they have all the information?

But they are not being given the information. It is about time they are.[81]

At the time Senator Ron Johnson delivered his remarks, I'd already been driven from government service three months before. But I was happy to see that somebody was standing up, telling the truth, and demanding answers. Unlike many Americans, I was not surprised to have this United States senator telling the world that even with his lofty elected position, he was subject to censorship by the COVID gods. Senator Ron Johnson from Wisconsin has emerged as one of the bright lights in the US Senate (along with some other senators and house members such as Marjorie Taylor Greene, Jim Jordan, Steve Scalise, and Lauren Boebert) and has really fought to get robust scientific information out to the public and has fought the censorship I and other scientists have experienced.

For these reasons and others, I did not have a positive view of Dr. Fauci from the time I began my work with HHS to support the COVID-19 commission through my tenure and afterward. I considered him a basic bench scientist who had somehow gotten into administration and, since that time, had been failing upward, gaining more and more power although there were few things he'd done right, which could account for his increasing authority. This is my opinion, and I have to be as honest with my view on his technical capacity, especially in the arena of public health and epidemiology.

I was also puzzled by his shifting stances on many important issues, such as proclaiming masks were not effective in a pandemic (which the evidence suggested was true) to doing an about-face and saying we all needed to be wearing masks as much as possible.

Dr. Fauci was also the highest paid member of the United States government at $434,312 in 2020,[82] (more than even President Trump), but whenever he took the podium to speak, I felt as if I was quickly identifying two to three completely wrong assertions of fact.

How was this man the most highly paid member of the United States government?

It boggled the mind.

As far as I could determine and, as alluded to earlier, it seemed Dr. Fauci had been a decent bench scientist, but quickly moved to the administrative track without really understanding public health or epidemiology. Perhaps it was a harsh assessment, but Dr. Fauci reminded me of many people I had met in academia, who hide in the university, only having the skills to write research grants, but never producing anything that is genuinely helpful to society. Many of the people I met in academia had very few practical "real-world" skills, such as understanding how to run a business, make a deadline, or balance a budget. They live in a comfortable, insular world they have created for themselves. Their goal in life is not to fix anything, but to simply exist, write applications for research grants, and collect their paycheck until their pension vested.

I considered Dr. Fauci to be politically incompetent. He'd hurt the gay community, and with COVID-19 it seemed he was auditioning to be the most famous person in America. Yet Trump, in my opinion, while on track to be the greatest president the United States has had (as I saw it in January 2020, based on his accomplishments), made two of the greatest mistakes historically by allowing Dr. Fauci and Dr. Birx to lead the lockdown and school closure response. This was a catastrophic failure, as it cost thousands of lives. The lockdown and school closure collateral harms and deaths lay at the feet of Dr. Fauci and Dr. Birx, who failed in providing accurate and optimal counsel to the president of the United States. They undercut him.

Early in my tenure at HHS in support of the task force I was told I simply had to let Dr. Fauci say and do what he wanted. He was the undisputed emperor of the federal government, like the dwarf tyrant Lord Maximus Farquaad in the *Shrek* movies. Because if we interfered in any way with what Dr. Fauci wanted to say, there would be leaks to CNN or MSNBC about us at HHS. This was a very serious matter as it could paralyze a president and his or her administration, and Dr. Fauci and his team routinely threatened us at HHS and especially the communications arm that they would leak lies and mistruths. The objective was to leak and get the complicit media to drive the narrative. In an election year, you could well imagine this was a rate-limiting step (a.k.a., the slowest step that occurs for a chemical reaction) as far as the administration was concerned, and Dr. Fauci et al. knew this and played it well, like the sword of Damocles. And we could pretty much count on the media simply parroting the leaks, and never finding out if there was any substance behind the claims. The group of people with whom I interacted referred to him as "Fauci, the diva" because of his constant demands and the fact he was inept, incompetent, and thoroughly corrupt.

In addition to being a diva, as we've established, Dr. Fauci had mismanaged the HIV/AIDS crisis, along with a number of other chronic health conditions, as well as the opioid epidemic that had exploded in the United States over the last thirty years. This is from a research paper in 2020 on the health status of our country:

> Approximately 47 percent of the U.S. population, 150 million Americans, suffered from at least one chronic disease, as of 2014. Almost 30 million Americans are living with five or more chronic diseases. The risk and prevalence of chronic disease grows as individuals age. Approximately 27 percent of children in the United States suffer from a chronic condition, while about 6 percent of children have more than one chronic condition. In contrast, around 60 percent of adults suffer from at least one chronic condition, while 42 percent suffer from multiple conditions. Among those 60 or older, at least 80 percent have one chronic illness and 50 percent have two. These ailments account for 70 percent of all deaths in America, killing more than 1.7 million people each year.[83]

Dr. Fauci didn't seem to know which policies might improve the health of Americans, but he did know how to use the power of the United States government to get the rest of the world to follow along. As it was explained to me during my time working at HHS in support of the COVID task force, the economic power of the United States was quickly deployed to get the other countries of the world to follow suit. Many of the poor and developing nations rely heavily on US currency transfers through Western Union and Moneygram, sent by citizens of those countries working in America and sending remittances back to their relatives.

Even the wealthier countries of the world feared the economic power of the United States, as even the hint of trade restrictions or transfer restrictions would quickly bring hesitant nations to heel.

Although I am a patriot to my core, I do not think the United States should have this level of influence over other nations and, in my opinion, I always felt President Trump would have not done this. Though, with time, I began to realize that the Republicans in Name Only (RINOs) and the Deep State bureaucrats were well capable.

* * *

Although I considered him just as corrupt as Dr. Fauci, Dr. Francis Collins, the longtime head of the National Institutes of Health (NIH), was a striking contrast to the mercurial Fauci who always needed to be in the spotlight.

Dr. Collins is a tall man, silver-haired, slender, gives off the air of a friendly country doctor, plays the guitar in his spare time, and makes much of the fact he's a Christian. But he was just as ruthless as Dr. Fauci. It was an especially good disguise for fooling the public, but it didn't fool me. Moreover, it was strongly believed that it was Dr. Collins who was the lead in the dynamic NIH/NIAID duo.

I like to say Dr. Collins was the silent partner to Dr. Fauci in his incompetent, mad scientist schemes.

When the documents came out showing that Dr. Fauci and his agency were trying to evade US prohibitions on gain of function (GoF) research by sending it to the lab in Wuhan, China, Collins's name was on those documents as well.[84]

But by staying away from the cameras, Dr. Collins was able to avoid a great deal of the media scrutiny. I've learned that you can be corrupt in Washington, DC, and, even if you get ambushed by an actual reporter and look questionable or like a criminal, if the attention lasts for just a day or two, you're probably going to be fine. It's only when some group decides to launch a sustained attack on you that you might be in real trouble.

However, sometimes the mask can slip, and the public can see a person's true face, as happened in an interview with Fox News's Brett Baier in December 2021 as Dr. Collins was retiring after twelve years at the helm of the National Institutes of Health:

> Outgoing National Institutes of Health Director Dr. Francis Collins took aim at the makers of "The Great Barrington Declaration," refusing to step down from calling them "fringe epidemiologists" while arguing "hundreds of thousands" would have died of COVID-19 if the country followed their advice.
>
> Collins told Fox News Host Brett Baier Sunday that he was "not going to apologize" for comments Friday in front of the House Select Subcommittee on the Coronavirus Crisis in which he called advocates of herd immunity "fringe epidemiologists," arguing that "hundreds of thousands of people would have died if we had followed that strategy."
>
> "I did write that, and I will stand by that," Collins said. "Basically, these fringe epidemiologists who really did not have the credentials to be making such a grand sweeping statement, were just saying let the virus run through the population and eventually then everybody would have had it and everything will be okay."[85]

The article goes on to note that, despite Dr. Collins' admonition that hundreds of thousands would have died, the death toll as of December 2021 was over 800,000 lives lost.[86] When faced with the evidence of his unprofessional conduct with Dr. Fauci by going after those scientists with a different opinion, Dr. Collins doubled down on his actions.

> Yet Collins called for a "quick and devastating published takedown" of the three experts pushing for a herd immunity strategy, which The Great Barrington Declaration largely pushes for.
>
> "So I'm sorry I was opposed to that, I still am, and I'm not going to apologize for it," Collins said. [Isn't that a non-sequitur?]
>
> Collins, who made the Fox News appearance during his last day on the job as NIH director, also pushed back against rumors that he was stepping aside because of his agency's alleged involvement in gain-of-function research in China, arguing that the theory COVID-19 leaked from a Chinese lab was a "huge distraction."
>
> "There is no evidence really to say. Most of the scientific community, myself included, think that is a possibility, but far more likely, this was a natural way in which a virus left a bat, maybe traveled through some other species and got to humans," Collins said.[87]

Despite what Dr. Collins claims, tens of thousands of animals have been tested, and scientists have not been able to find an infected animal that might have served as a reservoir for the virus, as detailed next. However, a report from *MIT Technology Review* in March 2021, a year after the United States locked down, found no evidence that the virus was a result of "spillover" zoonotic transfer, whereby it jumped from an animal to human beings. This is how they responded to the question of what was known and what remained unknown.

> What the researchers know so far is that a coronavirus very similar to some found in horseshoe bats made the jump into humans, appeared in the Chinese city of Wuhan by December 2019, and from there ignited the biggest health calamity of the 21st century.
>
> We also know they haven't found the critical detail: if it was in fact a virus with an origin in horseshoe bats, how did it make its way into humans from creatures living hundreds of miles away in remote caves?[88]

Aside from finding an effective treatment for the virus, there was probably no more critical question than determining the origin of the virus. But even after a year of rampaging through our world, the researchers could find no evidence of an intermediate animal host.

> This time, though, the intermediate-host hypothesis has one big problem. More than a year after covid-19 began, no food animal has been identified as a reservoir for the pandemic virus. That's despite efforts by China to test tens of thousands of animals, including pigs, goats, and geese, according to Liang Wannian, who leads the Chinese side of the research team. No one has found a "direct progenitor" of the virus, he says, and therefore the pandemic "remains an unsolved mystery."
>
> It's important to know how the pandemic started, because after killing more than 2.5 million people and causing trillions of dollars in economic losses, it's not over. The virus may well be establishing itself in new species, like wild rabbits or even house pets. Learning how the pandemic began could help health experts avert the next one, or at least react more swiftly.[89]

So much of what happened in the investigation of the origins of SARS-CoV-2 reeked of a coverup, both by the Chinese Communist Party (CCP), and their lackeys in the United States scientific community like Dr. Collins and Dr. Fauci. If you weren't a part of the scientific community who was working on this dangerous gain of function research (NIH, NIAID, EcoHealth Alliance [Dr. Peter Daszak], University of North Carolina, Chapel Hill [Dr. Ralph Baric], Wuhan Institute of Virology [WIV], Dr. Fauci, Dr. Collins etc.), it was clear where your suspicion should be directed.

> But some fear that all the research into bat viruses may have backfired in a shocking way. These people point to a striking coincidence: the Wuhan Institute of Virology, the world epicenter of research on dangerous SARS-like bat coronaviruses, to which SARS-CoV-2 is related, is in the same city where the pandemic first broke loose. They suspect that covid-19 is the result of an accidental leak from the lab.
>
> "It's possible they caused a pandemic they were intending to prevent," says Matthew Pottinger, a former deputy national security advisor at the White House. Pottinger, who was a journalist working in China during the original SARS outbreak, believes it is "very much possible that it did emerge from the laboratory" and that the Chinese government is loath to admit it.

Pottinger says that is why Beijing's joint research with the WHO "is completely insufficient as far as a credible investigation."[90]

As the MIT article details, at the beginning of the SARS-CoV-2 outbreak it might have made sense to link it to the Wuhan Wholesale Seafood Market. In the original SARS-CoV-1 outbreak, "it quickly became clear that chefs and people handling animals were the first cases. More of them had antibodies to the virus."[91] Then the MIT article delivered the kicker.

> The hunt this time is fundamentally different. A likely origin for covid-19 is already known: it's very close to known bat viruses. Even before the outbreak started, the Wuhan Institute had studied one whose genetic code is 96% identical to SARS-CoV-2. That's as good a match as the "missing link" found for the original SARS.
>
> That means the burning question now isn't so much the deep origin of the virus as how such a pathogen would have ended up in the city of Wuhan . . .
>
> So far, there is no evidence the outbreak went undetected elsewhere before the Wuhan cases. Genetic evidence also narrows the chance that the virus was spreading much earlier. Because of how the germ has accumulated mutations with time, it's possible to estimate when it first started spreading between people. That data, too, points to a start date of late 2019.[92]

And yet Dr. Collins wants you to believe that all of these questions are simply a distraction from his really important work.

It's often said that quiet people are sometimes the most dangerous, and I believe that warning definitely applies to Dr. Francis Collins. Dr. Collins must be investigated along with Dr. Fauci and others, as to their roles in GoF research, with strong accountability if it is shown that their actions were illicit and caused harms and deaths needlessly.

* * *

While Dr. Collins deftly downplayed the idea that SARS-CoV-2 was leaked from the Wuhan Institute of Virology, there was another equally heinous crime he committed, and that was the unprovoked and unwarranted attack on three highly qualified scientists, Jay Bhattacharya of Stanford University, Sunetra Gupta of Oxford University, and Martin Kulldorff of Harvard University, authors of the Great Barrington Declaration

What's genuinely remarkable about these three scientists is that in addition to being accomplished in their fields, they have that rare type of intellect that allows them to be similarly successful in a different field. One might argue there is the relatively common brilliance of being able to succeed in a narrow field of specialty, but there is something unique about the intellectual abilities of those who can rise to the top in multiple areas of intellectual pursuit.

I believe such individuals are more likely to have a broader view of the world, and are less likely to make, or suggest, plans that might result in catastrophe. As an example, here's the bio for Jay Bhattacharya of Stanford.

> Jay Bhattacharya is a Professor of Medicine at Stanford University. He is a research associate at the National Bureau of Economics Research, a senior fellow at the Stanford Institute for Economic Policy Research, and at the Freeman Spogli Institute. He holds courtesy appointments as Professor in Economics and in Health Research and Policy. He also directs the Stanford Center on the Demography of Health and Aging. Dr. Bhattacharya's research focuses on the economics of health care around the world with a particular emphasis on the health and well-being of vulnerable populations. Dr. Bhattacharya's peer-reviewed research has been published in economics, statistics, legal, medical, public health, and health policy journals. He holds an MD and PhD in economics from Stanford University.[93]

Bhattacharya is smart enough to get a MD from Stanford Medical School. But that wasn't enough for him. He went back to Stanford University and got a PhD in Economics, studying under brilliant economists like future Nobel Prize winner Professor Peter J. Klenow. In addition, he's published in economic, legal, medical, public health, and public policy journals. But Dr. Collins takes this to mean that brilliance displayed across multiple fields means that he can't be good at any one of them.

A similar brilliance across multiple fields has been demonstrated by Professor Sunetra Gupta of Oxford University. This is from her biography on the Merton College, Oxford University website:

> Professor Sunetra Gupta is a novelist, essayist, and scientist. She is currently Professor of Theoretical Epidemiology at Oxford University's Department of Zoology and a Supernumerary Fellow at Merton College.
>
> Born in Calcutta (now Kolkata), India, Sunetra graduated from Princeton university in 1987 and received her PhD from Imperial College,

London in 1992. She started her career at Merton in the following year as a Junior Research Fellow in Zoology. Her research on infectious disease agents that are responsible for malaria, HIV, influenza, bacterial meningitis, and pneumonia. Among her many achievements, she has invented a new method of producing a universal influenza vaccine which has been licensed by Blue Water Vaccines in the USA. She was awarded the 2007 Scientific Medal by the Zoological Society of London and the 2009 Royal Society Rosalind Franklin Award.

Sunetra is also a novelist, having written five works of fiction, and is an accomplished translator of the poetry of the Bengali polymath Rabindranath Tagore. Her books have been awarded the Sahitya Akademi Award and the Southern Arts Literature Prize, shortlisted for the Crossword Award, and longlisted for the Orange Prize and the DSC Prize for South Asian Literature.[94]

Unlike Dr. Fauci or Dr. Collins, Dr. Gupta is an actual epidemiologist. In addition to her work in epidemiology (mainly an academic field entailing a good deal of reviewing the published and unpublished work of others), Dr. Gupta shows herself to be an unusually effective bench researcher and developer of new technologies, in addition to being an award-winning translator and novelist.

Dr. Fauci and Dr. Collins seem to believe the only brilliant people are those who have worked in government agencies for decades. The rest of the world may beg to differ.

And if there's somebody you'd like to be in charge of making sure that a new strategy was being rolled out safely, you could hardly ask for somebody more qualified than Dr. Martin Kulldorf:

Martin Kulldorff, PhD, is a biostatistician, an epidemiologist and a Professor in the Department of Medicine at Harvard University (on leave), a senior scholar at the Brownstone Institute and a fellow at Hillsdale College's Academy for Science and Freedom. Dr. Kulldorf's research centers on developing and applying new disease surveillance methods for post-market drug and vaccine safety surveillance and for the early detection and monitoring of infectious diseases outbreaks. In October 2020, he coauthored the Great Barrington Declaration, advocating for a pandemic strategy of focused protection instead of lockdowns.

Dr. Kulldorf has developed new sequential statistical methods for near real-time post-market drug and vaccine safety surveillance, where the purpose is to use weekly or other frequent data feeds to find potential safety problems as soon as possible. He has also developed tree-based scan statistic data mining methods for post market drug and vaccine safety surveillance. Keeping the outcome definitions flexible, the methods simultaneously evaluates thousands of potential adverse events and groups of related events, adjusting for the multiple testing inherent in such an approach. These methods are used by the FDA and CDC to monitor drug and vaccine safety.[95]

Again, unlike Anthony Fauci or Francis Collins, Dr. Kulldorff is an actual epidemiologist. However, apparently because he has additional skills, such as building safety systems to monitor drugs and vaccines, he's a "fringe epidemiologist." It's easy to imagine that any of these three individuals would have done a way better job than Dr. Fauci and Dr. Collins during the COVID-19 crisis, while also maintaining the public's trust in the integrity of the scientific establishment.

And what about the Great Barrington Declaration, which Dr. Fauci and Dr. Collins wanted to attack so viciously? How many people have actually read it? Is it so far from the mainstream of acceptable scientific thought that we must renounce it as if it was some crazy idea like spraying Windex on a cut? I'll let you be the judge.

However, I'd also like you to be aware of the timing. I was hired in May 2020 and forced out on September 25, 2020. During those months, I was part of a small group, which included Scott Atlas, Michael Caputo, Peter Navarro, and others, who were advocating for a limited lockdown, a focused approach to management of the emergency using an age-risk stratified approach, and the provision of genuine health advice, which might blunt the worst parts of the outbreak and allow the economic life of the country to continue. The Great Barrington Declaration was released on October 4, 2020, less than two weeks after I was forced to resign.

As infectious disease epidemiologists and public health scientists we have grave concerns about the damaging physical and mental health impacts of the prevailing COVID-19 policies, an approach we call Focused Protection.

Coming from both the left and the right, and around the world, we have devoted our careers to protecting people. Current lockdown policies are producing devastating effects on short and long-term public health. The results (to name a few) include lower childhood vaccination rates, worsening cardiovascular disease outcomes, fewer cancer screenings and deteriorating mental health—leading to greater excess mortality in years to come, with the working class and younger members of society carrying the heaviest burden. Keeping students out of school is a grave injustice.

Keeping these measures in place until a vaccine is available will cause irreparable damage with the underprivileged disproportionally harmed.

Fortunately, our understanding of the virus is growing. We know that vulnerability to death from COVID-19 is more than a thousand-fold higher in the old and infirm than the young. Indeed, for children, COVID-19 is less dangerous than many other harms, including influenza.

As immunity builds in the population, the risk of infection to all— including the vulnerable—falls. We know that all populations will eventually reach herd immunity—i.e. the point at which the rate of new infections is stable—and this can be assisted (but is not dependent upon) a vaccine. Our goal therefore should be to minimize mortality and social harm until we reach herd immunity.

The most compassionate approach that balances the risk and benefits of reaching herd immunity, is to allow those who are at minimal risk of death to live their lives normally to build up immunity to the virus through natural infection, while better protecting those who are at highest risk. We call this Focused Protection.

Adopting measures to protect the vulnerable should be the central aim of public health responses to COVID-19. By way of example, nursing homes should use staff with acquired immunity and perform frequent testing of staff and all visitors. Staff rotation should be minimized. Retired people living at home should have groceries and other essentials delivered to their home. When possible, they should meet family members outside rather than inside. A comprehensive and detailed list of measures, including approaches to multi-generational households, can be implemented, and is well within the scope and capability of public health professionals.

Those who are not vulnerable should immediately be allowed to resume life as normal. Simple hygiene measures, such as hand washing and staying home when sick should be practiced by everyone to reduce the herd immunity threshold. Schools and universities should be open for in-person teaching. Extracurricular activities, such as sports, should be resumed. Young low-risk adults should work normally, rather than from home. Restaurants and other businesses should open. Arts, music, sport and other cultural activities should resume. People who are more at risk may participate if they wish, while society as a whole enjoys the protection conferred upon the vulnerable by those who have built up herd immunity.[96]

The Great Barrington Declaration said in public what people like me, Scott Atlas, Michael Caputo, and Peter Navarro had been saying in private to the members of the COVID-19 task force, as well as to the White House and President Trump, for months.

The schools should have been opened and should have never been closed.

The economy and society should have been opened and should have never been shut down, save but for maybe two weeks at most in the initial stages.

Businesses should have never been closed, and employees should have never been laid off.

People should have been out at cultural events or playing sports.

The churches should have been welcoming parishioners and members of their congregations and not have been closed.

The able-bodied, even up to the age of seventy-five, if they didn't have any significant pre-existing conditions, should have been involved in society in ways of their choosing.

For those over the age of seventy-five, or those with significant pre-existing conditions, they should have remained at home, having groceries delivered to them, and limiting contact with the outside world.

If somebody was living in a multi-generational household with older people, appropriate guidelines could have been established by taking common sense safety precautions.

People should have been given accurate information, but then it was up to them to determine the level of risk they wanted to take.

No mandates of any kind. None.

Why was it necessary in October 2020 for Dr. Fauci and Dr. Collins to launch such a vicious and illegitimate attack against Bhattacharya, Gupta, and Kulldorff and their Great Barrington Declaration?

The election of November 8, 2020 is my answer.

If the principles of the Great Barrington Declaration had been implemented in October 2020, there would have been an explosion of joy in this country unlike anything experienced since the end of World War II.

Trump would have been hailed as the victor over COVID-19, with the summer of fear giving way to a holiday season of hope.

However, that didn't happen.

Let me tell you how this small group of individuals changed history.

CHAPTER FOUR

Trump in the Crisis—A Tale of Two Januarys

I'm aware that readers of this book are likely to be sharply divided about this following chapter. For some, it will be a vindication of things they already believed (but hopefully providing greater detail to support such beliefs), while others will accuse me of engaging in wild speculation regarding some of the leading figures of our government, including the president of the United States.

I understand both reactions.

However, I was there, called to and working at HHS, having direct contact with members of the task force, having regular contact with various scientists and officials from CDC, NIH, FDA, Moderna, Pfizer, etc., and having people come to me routinely and confidentially to express their own views and their frustrations with the lockdowns and the vaccine development under OWS, their own fears of wrong and risks, as someone playing a supportive role to the COVID-19 task force from HHS—all this during some of the most critical times in our nation's history. I may not be able to tell you exactly what is true, but I can probably get you closer to the truth than just about anybody in the media talking about the situation today.

And besides, other members of the task force, like Dr. Deborah Birx and Dr. Scott Atlas, have published their own books, so why shouldn't I have the right to tell my story and my interpretation of those events?

I think I have a perspective you are not likely to hear anyplace else.

I call it, "A Tale of Two Januarys."

* * *

When I was at HHS and playing my role on the task force, I was told about January 2020 meetings that no doubt were involving President Trump's team at the White House and those ramping up the re-election push.

The polling (internal and otherwise) was excellent news for the president, predicting he was likely to win in a landslide with somewhere between forty to forty-five states, and as a result Trump was in an upbeat mood. His initial rallies were attended by tens of thousands inside and outside of the halls. The economy was doing great and employment among all Americans—men women, Black, white, Asian, and Hispanic—was at an all-time high. He had run on a campaign of being an iconoclastic, brash entrepreneur (not as an angel), and while his combative tweets had horrified the press and his democratic opponents, it had thrilled many in the public (democrats, republicans, and independents) who believed a politician was finally revealing the corruption in our government that they'd long suspected. Finally, they had a person who would take on the establishment and not be afraid to represent their fears, their wants, and their beliefs. Who would fight for them. Again, the preliminary January 2020 polls were fantastic, for the public knew based on what he'd accomplished, without the help of even those in his own party, that the decades ahead looked very bright!

After the panic of the COVID-19 crisis it can be difficult to put yourself back in the frame of mind of how the world looked as the curtain went down on 2019. Let me help you with the benefit of an article published on December 25, 2019, looking back on the highlights of the previous year.

The ongoing government shutdown becomes the longest in U.S. history on **Jan. 12** when it reaches 22 days; 800,000 employees are left unpaid. On **Jan. 25,** President Trump agrees to temporarily end the shutdown by supporting a deal to find federal agencies for three weeks.

The U.S. Justice Department charges Chinese technology firm Huawei with multiple counts of fraud on **Jan. 28** . . .

President Trump declares a national emergency on **Feb. 15** for access to funds for his proposed border wall. He issues the first veto of his presidency on **March 15**, striking down a Senate resolution to end his national emergency declaration . . .

On **March 24**, U.S. Attorney William Barr publishes a four-page summary of Special Counsel Robert Mueller's report into U.S. President Donald Trump's 2016 election campaign. On **April 18**, the full 448-page report is released in redacted form. [The report clears Trump officials as well as declaring that not a single American, "wittingly or unwittingly" colluded with Russian disinformation efforts.] . . .

The New York Times publishes newly obtained tax information on **May 8**, which reveals that from 1985 to 1994, Donald Trump lost $1.7 billion from his various businesses . . .

President Trump becomes, on **June 30,** the first sitting U.S. President to cross the Korean Demilitarized Zone and enter North Korea . . .

On **July 6**, billionaire registered sex offender Jeffrey Epstein is arrested on federal and state charges of sex trafficking. Epstein is found dead of apparent suicide in his cell on **Aug. 10** . . .

A whistleblower files a complaint on **Aug. 12**, which alleges that the President of the United States used the power of his office to solicit interference from a foreign country [Ukraine] in the 2020 U.S. election. Michael Atkinson, the inspector general for the intelligence community, deems the complaint an "urgent concern" and "credible" on **Aug. 26** . . .

President Donald Trump reveals on **Sept. 7** that he has cancelled planned peace talks with the Taliban at Camp David . . .

On **Sept. 24**, U.S. Speaker of the House Nancy Pelosi announces the start of a formal impeachment inquiry against President Donald Trump. On **Oct. 31**, the House of representatives votes 232-196 in favor of formally proceeding with an impeachment inquiry against the president . . .

President Trump announces on **Oct. 27** that the leader of ISIS, Abu Bakr al-Baghdadi, was killed in a U.S. Special Forces operation . . .

Public impeachment hearings against President Trump begin in the House of Representatives on **Nov. 13**. On **December 13**, democrats on the House Judiciary Committee approve abuse of power and obstruction of Congress charges against the president who becomes the fourth U.S. president in history to face impeachment. On **Dec. 18**, the House votes to forward the two articles of impeachment to the Senate . . .

The Inspector General of the Department of Justice, Michael Horowitz, issues a report on **Dec. 9** that concludes that the Federal Bureau of Investigation's (FBI) inquiry into the 2016 Trump campaign was legally justified and conducted without political bias albeit with several procedural errors most notably in the Foreign Intelligence Surveillance Court (FISA) warrants.[97]

Do you feel like you have a little better handle now on the events of 2019? Trump was fighting with the Deep State bureaucracy on spending, forcing the longest government shutdown in history, charging China with the theft of US technology, making sure his border wall got built, getting cleared of the charges of Russian collusion in the 2016 election, walking into North Korea and shaking hands with Kim Jong-Un (and thus dramatically lowering tensions in the world's most volatile region), making sure Jeffrey Epstein got arrested, and killing the leader of ISIS, Abu Bakar al-Baghdadi.

If you were President Trump, you might be forgiven for thinking you'd had a pretty good year, especially since in that time he'd destroyed the Russian collusion hoax, the hoax about getting peed on by Russian hookers in a Moscow hotel room in a "golden shower" (even though he's an admitted germaphobe), the Charlottesville "fine people" hoax, and the hoax Trump was such a philistine that he overfed the goldfish when he visited with Prime Minister Shinzo Abe of Japan in November 2017. [98] (Oh, the horror of an accusation of overfeeding goldfish! We must not let such monsters walk among us! Put him in a jail cell next to Jeffrey Epstein!)

This is how Steve Cortes described America on the brink of 2020 in a December 31, 2019 article in *Real Clear Politics*:

> America stands on the cusp of a new decade that promises to unfold as the new "Roaring Twenties." A review of President Trump's 2019 achievements, building on the successes of 2017 and 2018, provide context for the year and decade ahead, and reasons to expect a resounding Trump reelection next November. [99]

From that point, Cortes went on to describe what he thought of Trump's ten most important accomplishments of the year. He considered them to be Trump's record in creating jobs, broadening his campaign movement (including recent polling indicating nearly 40 percent support from Latinos), confronting China, negotiating important trade deals, his record on appointing constitutionalist judges, the "Remain in Mexico" policy for asylum-seekers, full exoneration of all claims of Russian collusion in the Mueller Report, the killing of ISIS leader, Abu Bakr al-Baghdadi, the astonishing 60 percent increase in natural gas exports, and 500,000 new factory jobs under Trump's tenure, as well as establishing the sixth branch of our military, the Space Force. [100]

As Cortes concluded in his article, looking at the bright dawn of 2020:

> These ten achievements build a foundation for our nation to flourish in the New Year. In addition, these accomplishments exhibit his leadership skills, in

spite of a near-totally obstructionist House of Representatives and a consistently biased media establishment. Such accomplishments make the president the prohibitive favorite to win reelection over an unimpressive Democratic presidential stable of candidates. Looking bigger picture, the first three years of the Trump presidency have established the policy framework and upward momentum for a truly amazing decade ahead—the new Roaring Twenties.[101]

Even if you're a partisan Democrat, I think it's difficult to disagree with the proposition that, as 2020 dawned, President Trump looked to be in an excellent position to win re-election. The view in Washington, DC, I was told, was that Trump was unstoppable electorally, and even the Democrats were conceding that.

At the end of President Trump's first term, the White House published a fifty-eight-page list of his accomplishments. For reasons of brevity, I quote from just the opening of the document:

Unprecedented Economic Boom

Before the China Virus invaded our shores, we built the world's most prosperous economy.

- America gained 7 million new jobs—more than three times the government experts' projections.
- Middle-class family income increased nearly $6,000—more than five times the gains during the previous administration.
- The unemployment rate reached 3.5 percent, the lowest in half a century.
- Achieved 40 months in a row with more job openings than hirings.
- More Americans reported being employed than ever before—nearly 160 million.
- Jobless claims hit a nearly 50-year low.
- The number of people claiming unemployment insurance as a share of the population hit its lowest on record.
- Incomes rose in every single metro area in the United States for the first time in nearly 3 decades.

Delivered a future of greater promise and opportunity for citizens of all backgrounds.

- Unemployment rates for African Americans, Hispanic Americans, Asian Americans, veterans, individuals with disabilities, and those without a high school diploma all reached record lows.
- Unemployment for women hit its lowest rate in nearly 70 years.
- Lifted nearly 7 million people off food stamps.
- Poverty rates for African Americans and Hispanic Americans reached record lows.
- Income inequality fell for two straight years, and by the largest amount in over a decade.
- The bottom 50 percent of households saw a 40 percent increase in net worth.
- Wages rose fastest for low-income and blue-collar workers—a 16 percent pay increase.
- African American homeownership increased from 41.7 percent to 46.4 percent.

Brought jobs, factories, and industries back to USA.
- Created more than 1.2 million manufacturing and construction jobs.
- Put in place policies to bring back supply chains from overseas.
- Small business optimism broke a 35-year record in 2018.

Hit record stock market numbers and record 401Ks.
- The DOW closed above 20,000 for the first time in 2017 and topped 30,000 in 2020.
- The S&P 500 and NASDAQ have repeatedly notched record highs.[102]

It was an impressive list, and there were fifty-six more pages of accomplishments that I did not include. But then came COVID-19, and by January 31, 2020, Trump was already taking action.

The Trump Administration declared a public health emergency in the U.S. Friday in response to the global coronavirus outbreak.

"Today President Trump took decisive action to minimize the risk of novel coronavirus in the United States," said U.S. Health and Human Services Secretary Alex Azar at a White House press conference . . .

The declaration of a public health emergency—which will become effective Sunday at 5 p.m. ET—enables the government to take temporary measures to contain the spread of the virus, which has been confirmed in seven people . . .

In addition, the U.S. is temporarily suspending entry of most travelers arriving from China, or who have recently been in China, if they are not U.S. citizens.[103]

With seven confirmed cases in the United States, Trump took the boldest action of any US president in history to restrict travel from a country with a viral outbreak.

Alas, it wouldn't be enough.

As the *New York Times* reported on March 12, 2020, four days before the start of the lockdowns, this was starting to cast a pall over Trump's reelection chances:

President Trump faces the biggest challenge yet to his prospects of being re-elected, with his advisers' two major assumptions for the campaign—a booming economy and an opponent easily vilified as too far left—quickly evaporating.

After a year in which Mr. Trump had told voters that they must support his re-election or risk watching the economy decline, the stock market is reeling and economists are warning that a recession could be on the horizon because of the worsening spread of the coronavirus.

And instead of elevating Senator Bernie Sanders of Vermont, as Mr. Trump made clear was his hope, Democrats have suddenly and decisively swung from a flirtation with socialism to former Vice President Joseph R. Biden, Jr., who has run a primary campaign centered on a return to normalcy.[104]

What was I doing at this time? I was watching on in admiration at the many successes of Trump for, in my mind, while I did support him and he did deliver, it really did not matter to me if the successful president was Democrat or Republican. I identify as an independent and more libertarian yet will admit I hold conservative views. You can also say I am someone who had some liberal leanings in the past yet became aligned more right as time passed and I saw the failures of the liberal democrat logic and the damage it does. The mediocrity and dependency that it pushes that is always crushing.

My belief was always that we rally behind a president who would do good for America and help all of America's people, especially the more vulnerable sectors. Trump was accomplishing that in spades. I was busy with my work, having already stopped my job with the Infectious Disease Society of America with a strong purposeful pivot to providing vetted research publications and my evidence-based skills to the WHO and PAHO, but most of the time I felt like I was sitting on the sidelines.

And it wasn't like I was making much of an impact at the WHO and PAHO, as they'd recently been praising how well China was doing with containing the virus. My role was as an evidence-based synthesis expert, yet I did not make decisions, in spite of the fact that I dealt with many WHO and PAHO folk officially. I balanced my work with many serious questions regarding what China was doing with the response they declared, and their role in the origins of the virus (i.e., was it made in a lab via manipulation?). This latter point outraged me. This is from an article in *Science* on March 2, 2020:

> Chinese hospitals overflowing with COVID-19 patients a few weeks ago now have empty beds. Trials of experimental drugs are having difficulty enrolling enough eligible patients. And the number of new cases reported each day has plummeted the past few weeks.
>
> These are some of the startling observations in a report released on 28 February from a mission organized by the World Health Organization (WHO) and the Chinese government that allowed 13 foreigners to join 12 Chinese scientists on a tour of five cities in China to study the safety of the COVID-19 epidemic and the effectiveness of the country's response. The findings surprised several of the visiting scientists. "I thought there was no way those numbers could be real," says epidemiologist Tim Eckmanns of the Robert Koch Institute, who was part of the mission.[105]

The numbers weren't real. China was lying, but my colleagues at the World Health Organization had bought it, hook, line, and sinker.

And what they were also selling us was a similar, Chinese-style lockdown.

> The question is now whether the world can take lessons from China's apparent success—and whether the massive lockdowns and electronic surveillance methods imposed by an authoritarian government would work in other countries. "When you spend 20, 30 years in this business it's like, "Seriously, you're going to try and change that with those tactics?" says Bruce Aylward, a Canadian WHO epidemiologist who led the international team and briefed journalists about its findings in Beijing and Geneva last week. "Hundreds of

thousands of people in China did not get COVID-19 because of this aggressive response."

"This report poses difficult questions for all countries currently considering their response to COVID-19," says Steven Riley, an epidemiologist at Imperial College London. "The joint mission was highly productive and gave a unique insight into China's efforts to stem the virus from spread within mainland China and globally," adds Lawrence Gostin, a global health scholar at Georgetown University.[106]

Even though I was at the WHO at the time, I didn't believe the rosy reports that were coming from our experts. China had lied about the outbreak at the beginning, they were probably (and, in my mind, likely) lying about its origins, and they were lying about how well they were controlling it. They were lying in a way to impact US policy and response.

It was around the time this report was released that Dr. Fauci seemed to change his mind about lockdowns.

This was also probably driven by a February 27, 2020 *New York Times* opinion article by Dr. Peter Daszak (whose EcoHealth Alliance group laundered a $3.7 million dollar grant from Fauci's agency to the Wuhan Institute of Virology to study bat coronaviruses), which argued that COVID-19 was the dreaded Disease X that everybody had been fearing.

> In early 2018, during a meeting at the World Health Organization in Geneva, a group of experts I belong to (the R&D) Blueprint) coined the term "Disease X": We were referring to the next pandemic, which would be caused by an unknown, novel pathogen that hadn't yet entered the human population. As the world stands today on the edge of the pandemic precipice, it's worth taking a moment to consider whether Covid-19 is the disease our group was warning about.
>
> Disease X, we said back then, would likely result from a virus originating in animals and would emerge somewhere on the planet where economic development drives people and wildlife together. Disease X would probably be confused with other diseases early in the outbreak and would spread quickly and silently; exploiting networks of human travel and trade, it would reach multiple countries and thwart containment. Disease X would have a mortality rate higher than a seasonal flu but would spread as easily as the flu.
>
> In a nutshell, Covid-19 is Disease X.[107]

We had the liars in China telling us they were doing a great job with the pandemic, and liars like Peter Daszak saying this was a civilization-altering pathogen, but not mentioning the likelihood that they they'd probably created SARS-CoV-2. They'd done it with GoF research where they were stitching together various virus and parts (creating chimeric viruses) and via inserting furin cleavage sites to the spike glycoprotein to test "spillover" and infectiousness as well as lethality. With what I knew of the lunacy at the time, I was arguing to anyone with ears that we were heading down a very dangerous road.

I felt like nobody was listening.

Boy was I wrong.

In April 2020 or so, around a month after the lockdowns had been instituted, I was watching television, bristling at what was being said on the news, sometimes muttering back at the stupid things being spouted, when the phone rang and my wife picked it up. I considered the COVID-19 task force to be a clown show, and I'm not a person who hides my emotions about anything.

After saying hello to the caller, my wife was quiet for a few minutes. I wondered if it was some solicitor with a really good pitch, because she usually hangs up rather quickly with one of those phone calls, but she didn't.

Instead, in a quiet voice she said, "Paul, it's someone saying they are talking from or on behalf of the White House and US government and they want to talk to you."

I took the phone from her and said, "This is Dr. Paul Alexander," not quite believing her.

The person speaking on behalf of the White House, sounding credible enough, identified himself and the nature of his call, and then said, "We know of some things you've said publicly and read some of the recent papers you've published on different issues. And that's gotten the attention of persons in the Oval Office. And they wanted to know if you'd be interested in joining the administration and having a seat at the table."

"Are you guys joking?" I asked, still not believing this was real.

The man on the other end of the line replied, "Well, no. We want to know if you want to have a seat at the table. We've done our study of you. And we realized you could be an asset to the administration. You're somebody who can be trusted, and we like your technical competence. And we want you to join us."

I think it's important for the reader to understand what I say next because it best explains my frame of mind as I began my work.

"Are you saying the task force can't be trusted?" I asked. I tend to be very blunt and brutally honest, especially in these types of situations.

There was a pause on the other end of the line. "The president wants to expand the table with people he trusts and who are optimally competent."

"So he doesn't trust them?" I asked, pressing the question again.

"I wouldn't put it that way," he replied. "He doesn't feel he's being fully, properly, and optimally served by the present members of the task force. We want other people in those meetings, giving their insight. If we're all in agreement, we're good. If not, well, we have to figure out the next steps. Will you join us?"

"Yes, while I still do not believe this conversation is true, and I am amazed yet yes, absolutely, I am privileged," I replied.

I was told initially that I would be recruited to join HHS, which was just across the street from the US Capitol building. My role would be to serve as the senior COVID pandemic advisor to work with HHS Assistant Secretary Michael Caputo. My job would involve straight line reporting to Caputo, dotted line reporting to HHS Secretary Azar, and dotted line reporting to the White House (a principal contact being the chief of staff direct to our offices), as well as to all the other managers and directors of the different agencies, as per requests or agreements by my direct boss(es). I was informed that all of the major agencies—CDC, FDA, the Department of Defense (DoD)—had sub-offices in my building, which was also where all the members of the task force worked on a daily basis besides going to the White House or over to congress for briefings or hearings.

When Dr. Deborah Birx released her book on the pandemic, it gave me a little deeper insight into what might have been going on behind the scenes that caused Trump to reach out to somebody like me, as well as to Dr. Atlas. As Birx recalled:

> No sooner had we convinced the Trump administration to implement our version of a two-week shutdown than I was trying to figure out how to extend it. Fifteen Days to Slow the Spread was a start, but I knew it would be just that. I didn't have the numbers in front of me yet to make the case for extending it longer, but I had two weeks to get them. However hard it had been to get the fifteen-day shutdown approved, getting another one would be more difficult by many orders of magnitude.[108]

There it is in Dr. Birx's own words. When they told you it was "Fifteen Days to Stop the Spread," they were lying to you. Yes, Dr. Fauci, Dr. Collins, and

Dr. Birx were lying to you. In fact, others on the larger task force were lying to you. They had longer lockdowns planned. They just didn't trust you to understand the situation. They were in for a penny, in for a pound. Dr. Birx continued with her description of how she manipulated Trump.

> The president walked in. He was dressed in more casual clothes than I was used to seeing him wearing in the Oval, a pair of slacks and a polo shirt.
>
> The vice president looked at me and signaled me to speak.
>
> Despite my nerves, I plunged right into the deep end. Opening with my PowerPoint graphics, I said, "Mr. President, we need to take additional actions immediately. I'm recommending that we extend the Slow the Spread measures by thirty days."
>
> "What will happen if we don't do the thirty days?" he asked. He had cut to the chase.
>
> I paused for a second, then decided to hit close to home: "If we don't, I'm certain that we're going to fifty, a hundred, and potentially a *thousand* Elmhurst Hospitals. That means more trucks outside those facilities. That means more bodies inside those trucks. We're going to see city after city looking like what New York does right now. It will only get worse."[109]

Trump agreed to the request, and Dr. Birx was pleased with the decision, writing of him, "Whatever my view of him as a politician or a person was immaterial. In this one instance, he had listened to the data, looked at the graphs and the evidence, and he had made only choice he could—and in doing so, he was helping us deliver a crucial message to the American people."[110]

But the honeymoon between Dr. Birx and President Trump didn't last, as she recounted.

> "We will never shut down the country again. Never." President Trump's tone was emphatic, edged with agitation. Furrowing his brow, he concentrated his full attention on me. His pupils narrowed into hardened points of anger.
>
> We were standing in the narrow five-by-eight space just outside the formal White House Briefing Room, which was crammed with hardworking communications staff. It was the first week of April 3, mere days after the president had announced the thirty day extension of the Slow the Spread campaign to the American public, and the ground had shifted suddenly and without warning.
>
> I felt the blood drain from my face and I shivered slightly.

> A moment later, I stepped into another press briefing, swept along by the frigid wake coming from the president's broad back. I experienced that unnerving sensation of having crested a hill too quickly while driving.[111]

This was consistent with the effort to bring me to HHS and onto the task force in a supportive role. Even if you did not see me, Caputo and I (and others) were never too far away in terms of the teleconferences, calls, etc. As I was told, President Trump didn't believe he was being well-served by the task force.

Dr. Birx was a curious choice to become the coordinator of the COVID-19 task force, and an article published by Michael Senger of the Brownstone Institute in July 2022 with the title, "The Talented Mr. Pottinger: The U.S. Intelligence Agent Who Pushed Lockdowns," provides something of an explanation. He wrote:

> I barely knew who Matt Pottinger was until I read that he'd appointed Deborah Birx as White House Coronavirus Response Coordinator in her bizarrely self-incriminating memoir *Silent Invasion*. There's little information about Pottinger online.
>
> Yet Pottinger is portrayed as a leading protagonist in three different pro-lockdown books on America's response to Covid-19: *The Plague Year* by the New Yorker's Lawrence Wright, *Nightmare Scenario* by the Washington Post's Yasmeen Abutaleb, and *Chaos Under Heaven* by the Washington Post's Josh Rogin. Pottinger's singularly outsized role in pushing for alarm, shutdowns, mandates, and science from China in the early months of Covid is extremely well-documented.[112]

Whether you're a detective or a scientist trying to understand what's hidden, the truth doesn't simply walk up to you and declare itself. Before my time in HHS supporting the COVID-19 task force and communications, many of my colleagues hinted at the intelligence agency origins of SARS-CoV-2. During my time on the task force I heard people on the force say such things, and, after my time, the evidence simply continued to pour in. As Senger (who I consider to be an excellent researcher and very bright and technically sound) detailed Pottinger's past:

> The son of leading Department of Justice official Stanley Pottinger, Matt Pottinger graduated with a degree in Chinese Studies in 1998 before going to work as a journalist in China for seven years, where he reported on topics

including the original SARS. In 2005, Pottinger unexpectedly left journalism and obtained an age waiver to join the US Marine Corps.

Over several tours in Iraq and Afghanistan, Pottinger became a decorated intelligence officer and met General Michael Flynn, who later appointed him to the National Security Council (NSC). Pottinger was originally in line to be China Director, but Flynn gave him the more senior job of Asia Director.

Despite being new to civilian government, Pottinger outlasted many others in Trump's White House. In September 2019, Pottinger was named Deputy National Security Advisor, second only to National Security Advisor Robert O'Brien.[113]

Pottinger has had an extremely interesting career path. Let's look at it with a little more of a jaded eye, shall we?

Forgive me if I take a more cynical view of the rise of Matt Pottinger.

How does a spook get to the top of the national security pyramid?

By pretending not to be a spook, but somebody who just accidentally found himself in a powerful position and did a great job.

First, there's the government history, starting with his father's work in the justice department. That's a real law and order background, isn't it? Boy, Matt sure comes from good, patriotic American stock, one might say.

Second, there's that degree in Chinese studies. Let's be fair. There are two types of people who get degrees in Chinese studies, genuine academics and spooks. That's pretty much the extent of it, right? Maybe there's a third category, somebody who wants to make money in China who thinks that's the way to do it.

Third, Matt ends up in China, as what? A journalist. Now there are two types of people who are journalists, actual journalists and spooks.

Fourth, "Pottinger unexpectedly left journalism and obtained an age waiver to join the US Marine Corps." What an interesting career path. I don't think that's anything his high school guidance counselor would have even contemplated. The pool of China scholars who subsequently become journalists and then obtain an age waiver to join the US Marine Corps is so shockingly small that I admit to being completely baffled by this turn of events and cannot even venture a guess as to what it might mean. There's just not enough data to come to a conclusion.

Fifth, "Over several tours in Iraq and Afghanistan, Pottinger became a decorated intelligence officer and met General Michael Flynn, who later appointed him to the National Security Council (NSC)." Does it seem clear that Pottinger was probably a spook since his college days, his years as a "journalist" in China, that sudden switch to the US Marine Corps, and suddenly he "met" Michael Flynn, "who later appointed him to the National Security Council (NSC)"?

I'm sure that kind of thing happens to people all the time who meet General Flynn.

"Hey, who are you and what are you doing in the White House situation room?" the president asks an unfamiliar face during a secret conference over Chinese plans for Taiwan.

"I don't know," says the uncomfortable young man. "I was just serving drinks to General Flynn last night at a Georgetown party, then this morning some secret service agents showed up and told me I've been appointed to the National Security Council."

How did Pottinger make the connection with Dr. Deborah Birx? His wife, Yen (Duong) Pottinger, a refugee from Vietnam, used to work for Dr. Birx. As recounted in the book *Nightmare Scenario* by *Washington Post* writer Yasmeen Abutaleb:

> When she became head of the CDC's Division of Global HIV/AIDS, one of her subordinates was a bright virologist named Yen Duong, who developed a widely used HIV test while working at the agency. Duong would eventually marry a *Wall Street Journal* reporter turned marine named Matt Pottinger, a connection that would eventually bring Birx into Trump's orbit.[114]

And to give a more complete picture of Yen, I'll quote from her profile from the Emma Willard School, which celebrated her as a "Distinguished Alumna."

> For as long as she can remember, Yen Pottinger has had a deep curiosity about how viruses work. "I love viruses over organisms like bacteria and parasites," Yen says excitedly. "Viruses are very efficient. They get the job done. HIV, for instance, is especially brilliant—it attacks the cells that you use to attack the virus, and it integrates itself into your genome so once you're infected, you can get rid of it," she says. A worthy opponent for this modern-day science hero.[115]

That's Yen Pottinger, the virologist, who simply happened to be married to the deputy National Security advisor. Maybe it was just a fortuitous chain of events. Here's how Dr. Birx recounted the approach for her to join the COVID task force.

On January 28, after meeting with Erin Walsh to solidify the planning and schedule for the upcoming African Diplomatic Corps State Department meeting, I received a text from Yen Pottinger. Aside from being the wife of my friend Matt, the deputy national security advisor, Yen was also a former colleague at the CDC and a trusted friend and neighbor . . .

Yen knew I would be on the White House complex for my meeting with Erin Walsh and the text she sent me said that Matt had a "proposition" for me. She didn't know any of the details, but Matt had apologized for the short notice and said he hoped we could meet face-to-face. Yen arranged so I could meet him in the West Wing, and once we were both there, Matt got to the point quickly. He offered me the position of White House spokesperson on the virus.[116]

Dr. Birx turned Pottinger down for the position of White House spokesperson in late January 2020, but Yen kept urging Dr. Birx to take some position to help with the virus. And Dr. Birx kept providing Matt with her thoughts:

I was giving Matt a lot of ideas, and I was glad to be able to help him, and the country, but I had one lingering regret. I texted Yen: Every time I turn something down, I feel like I am making yours and your children's lives more difficult, as Matt has to work more. But I am trying to support behind the scenes.

She responded: "That's funny, because it's not true. Don't feel bad. Our lives will continue to be difficult until he finds a new job. He thinks you should take over Azar, Fauci, and Redfield's jobs, because you're such a better leader than they are. He has been underwhelmed thus far."[117]

In her own words, you have Dr. Birx being told that the deputy National Security advisor to Trump, and his brilliant virologist wife, were "underwhelmed" by the performance of Alex Azar, Anthony Fauci, and Robert Redfield. Those are not my words. They are my opinion, but that's even before I came on the scene.

As the crisis continued to deepen in February, Matt and Yen continued to press Dr. Birx to join the administration, saying they wanted to add her name to the shortlist for a coronavirus task force coordinator. She recalled:

On February 26, Matt called me expressing greater worry. He told me that every moment I delayed making a decision, I could potentially be costing American lives. Having served in the military for twenty-nine years, I knew that when a senior leader asks you to take on a mission for the American people, you commit to the mission. And you stay on mission 24/7 until it is completed or you are removed. That's what soldiers do. That's what anyone called to serve does. As simple as Yen and Matt's message was—that I was needed—it contained multitudes.

The next day, I phoned Matt to let him know he could add my name to the shortlist.

If they decided they needed me, I was in.[118]

It's a moving account similar to my own experience, except it was three months later and the White House had added Dr. Birx to the list of people they didn't respect.

And yet, somehow, Matt Pottinger seemed to glide through all of it, as if he was the "Teflon bureaucrat." Michael Senger's article described Pottinger's fate and future as follows:

> While many Trump administration officials have floundered since Trump left the White House, "things are going well for Pottinger," Vox gushed. "[T]hat subject matter expertise—plus the patina afforded by resigning on January 6—has helped Pottinger, a former journalist, expertly navigate the post-Trump landscape. He even emerged as the White House hero of the initial Covid-19 chaos in *New Yorker* writer Lawrence Wright's chronicle of *The Plague Year* . . . One reason that Matt Pottinger was welcomed back into the establishment is that, unlike some of Trump's unconventional appointees, he had already been a part of the elite."
>
> From the center-right to the center-left and the far right to the far left, it's tough to find anyone on the Beltway short of praise for Matt Pottinger. Everything about Pottinger is silky smooth. Between the lines of glowing coverage are not-so-subtle winks and nudges that he'd make an excellent candidate for higher office.[119]

It's almost spooky how this former marine could fight in the mud of the Washington, DC, swamp and yet emerge with such a spotless reputation.

Forgive me if I'm skeptical of the claim.

* * *

I understand the media likes to portray Trump as a shoot-from-the-hip guy, and, compared to most politicians, it's easy to understand why he might appear that way.

Trump invited all the members of the COVID task force (groups that I was involved with) to a reception at the Trump Hotel in Washington, DC. I was also in meetings that could allow me to form a decent workable opinion of Trump. I can also say, without specifics, that I have sat in on calls directly to the White House, to the Oval Office, and I can say that Trump was extremely understanding and empathetic to the challenges, losses, and pains of what the lockdowns were doing to business owners and to laid-off employees, and the harms accrued by children due to the school closures. Trump was always sympathetic and sought all answers to get society and schools urgently re-opened.

I think I've had enough interaction with Trump (to the extent that I may share) to form a pretty good picture of the man, both the positive and the negative. The overwhelming feeling you get from Trump is that of strength and power, that he wants you to feel safe and you have his complete attention, as if he's an attentive father or that brother who's with you a hundred percent, and, if there's a fight, he's got your back. He looks intently at the person who's speaking, shakes their hand in that old-time way of genuine friendliness, and his body language broadcasts that you have his complete attention. Trump commands the room, and he's very warm and inviting to the people in it. You get the feeling right away that he loves America, and is a huge defender of the flag and all of the rights and privileges that flow from the good governance.

The other consistent trait I observed in Trump is he was always in command of the evidence. The man does his homework. Unlike the criticisms I've heard of Reagan, Trump was always prepared. The other thing I don't think gets conveyed enough is the compassion of the man. He was not an aloof, distant, unconnected individual. When you are speaking with him, you get the sense he sees all of you. I'm not sure if I'm explaining it accurately. It's difficult to describe how this superpower of focused attention by a powerful leader is such an effective tool. However, it is, and you walk away feeling "safe." I recall talking to some women when I was in DC around mid-July 2020. Some of them were Republicans and some were Democrats, and they all told me they voted for him. I asked them why, as the campaign was heating up and the election was up in the air. One of the Democrats said something that helped me understand how Trump gained such a huge portion of the women's vote and helped me understand this way he connects. She said, with him in power, she felt "safe" and he made her feel that her young daughters were "safe" when he was the president on deck.

Trump liked to trust his gut, but that was just the beginning of making a decision for him. Like a good corporate CEO, he always liked to question why things were being done. That tendency was skewered in the press when, behind closed doors, he allegedly asked why we have NATO, as well as why we have nuclear weapons.

Those are the kind of questions a corporate CEO asks.

If there's an answer, then he expects the smart people in the room to provide the rationale. If they can't, maybe it's just a policy that has become so ingrained that people need to take a second look at it.

However, this is also where I believe Trump came to make bad decisions as president.

In the corporate world, he was generally well-served by the executives who worked for him. The consequences of success or failure of an approach could be determined in weeks or months. The executives under Trump in the business world didn't have their own spin-machine and separate levers of power, as Dr. Fauci and Dr. Collins did with the thirty-one billion dollars a year they could dole out in research grants to the scientific community, or direct access to media outlets that they could get to broadcast their version of reality in mere hours.

While I believe Trump understood this to be an unfair state of affairs, I don't know if he made the next intellectual leap, that this corrupt system was promoting policies that were actively harming the public, whether it be lockdowns, denigrating therapeutics such as hydroxychloroquine and ivermectin, or the promotion of unsafe vaccines. (I will cover the safety testing of vaccines in the next chapter.) It is as if he was very naïve in this regard. His medical advisors like Dr. Fauci, Dr. Birx, and others managed to sell him their version of the truth, which clearly was not based on good, sound science.

Trump didn't trust the task force, but wanted to make sure he understood what was really going on before making a consequential decision. He was taking a corporate approach to solving a problem, not understanding that the edifice of these government entities didn't just need a remodel but needed to be demolished and rebuilt from the ground up. Running successful businesses is what Trump excelled in. Unfortunately, you couldn't apply a business approach to an organization that didn't remotely resemble a business. To make the wheels of a government well-oiled, you had to understand the system, know how it works, and know how to negotiate with all sides. It was different than running a business to turn a profit.

I don't think I understood the immensity of it all at the beginning, either.

I was looking forward to being part of this historic moment in history and hoping I could help even in a small way and have some sort of impact.

* * *

The problem with being asked to work for the US government was that, at the time (April–May 2020), my wife and I were transitioning fully to the United States and actually had previously started living there, yet when I was asked to come to Washington, I was in Canada gathering some items and the border with the United States was fully closed. I was new to the United States and did not understand that, with my US residency, the closure did not apply to me. However, I was told that I should go to the border at a certain place, Buffalo, New York, at a certain time, and agents would be there to escort me across. It was as if I was going to be on the set of a spy movie.

You must remember that the entire US border with Canada was closed. No people and no shipments were going through. There was also lots of confusion at that time as to who could and could not get into the United States from Canada. It was like a ghost town, very different from the pre-COVID-restriction days when cars were lined up for hours waiting to cross into the United States. When I approached the border agents and told them my story as to why I was there, I'm sure they thought the impeccably dressed, olive-skinned man with a Caribbean accent must have been an escaped mental patient. *Yes, I'm here to cross the closed border because the United States government has asked me to join as part of the pandemic response to work at HHS and in support of the COVID-19 task force.*

Luckily, the US officials arrived a few minutes afterward and spoke to the agent and gate I had pulled up to, and, although the border agents looked at me somewhat dubiously, they were satisfied that the officials were representatives of Trump and the US government and that my reasons were legitimate, so they let me cross the border into the United States.

I sat initially in a vehicle on the US side of the border and was interviewed about a range of issues surrounding terrorism, my background, etc., and I realized I was being vetted. I still thought I was being tricked and that someone was playing a joke on me, for it was really as if I were in some sort of spy movie. From there I was driven to an office in Buffalo where I spent about three hours answering questions and calls to the White House. I continued to be asked the predictable terrorism questions, and about links

I might have to any individuals or groups. I was asked more detailed questions all about my background, prior work, and prior schooling. I assure you, outside of my work, I'm a very boring person. I like my work and spending time with my family. The vetting process took a few hours, and it seemed like I was being scrutinized the same way as when I delved into my past and provided personal and professional information as part of the US residency application. After the questioning was over, they also discussed what I would be doing in my role. I was to be one of the people Trump would trust at HHS and in any task force-related work to inform him and all those at the White House of the truth, and to back it up with hard science.

After the vetting, I proceeded to depart to Washington, DC. When we got to Washington, DC, it was completely surreal. I saw a few people on the streets, hidden behind their masks, but there was hardly any other activity. Washington was completely shut down. Only occasionally would you see a car going down the street. We could not check in to the hotel we had booked because we were told on arrival that it was being used to lodge health care workers who were dealing with COVID patients. It was the dead of night in the dark, eerily empty streets of DC, and we had to look for another hotel. The same situation occurred at approximately three other hotels that night until we managed to find one in fairly close proximity to the HHS building on Independence Avenue that I had to report to.

When I went to HHS for my orientation, one of the first things I was told was that the Deep State knew I was Trump's appointee and pick, and they'd try to thwart me, first by not completing the paper work for my appointment and then by not paying my salary and holding up my security clearances. For the first few weeks, I had to be escorted up the elevator to my office, daily, until I got security clearance, even though at this stage of the game, this was more of an administrative step. When talking about the Deep State, I think it's important to realize it's not the picture painted by the media of guys in crazy-looking military hats and uniforms, with guns, running around. It's the permanent bureaucracy of Washington, DC— the people at the agencies who believe in their deepest hearts that they run the country and we should thank them for it. I tried to push back by saying that I didn't want to be thought of as Trump's appointee to HHS in any way, as it was hurting my reception, but just an honest scientist using his expertise to help out. They told me that wasn't the way I was viewed.

Well, I thought, *I'll try to change that opinion.*

However, I also resolved myself to the idea that in some ways they might treat me as if I didn't exist, and I figured, *Well, I'll just take it day by day.*

The human resources paperwork and salary situation was critical, as I'd signed a year-long lease on a condo on New Jersey Avenue just down the street from the Capitol Building. The rent was about three thousand a month, and I had already made plans to have my furniture shipped. I also realized that if I didn't have any official status, how would they let me meet with people like Dr. Redfield, Dr. Birx, and Dr. Fauci? Eventually, I was granted status as a "scientific volunteer," whatever the hell that meant. It was a good thing the people I was interacting with generally didn't understand the bureaucratic hell I was enduring.

As I started to do my job, I would be in the important meetings with top officials and others, and then afterward my many bosses would ask me to come to their office and explain what had been presented. These people weren't scientists. They were smart in their own ways, and it was my job to turn the science into layman's language that they could understand.

Our line to the White House was usually Chief of Staff Mark Meadows (often via Caputo) or one of his subordinates, and in those early days I said to them, "I can't believe people would do something like this to me. I am a legal resident. I have a green card. But even though I'm being treated this way, I have the epidemiological evidence-based medicine skills that are needed. I was asked to serve America, I'm honored to do it, and if necessary, I will do it for free." It was my understanding that Chief of Staff Meadows was trying to get the Deep State bureaucrats to step back and complete my human resource paperwork and begin paying me. I was actually confident in that, for I knew of the calls Caputo was having with him and that this was a key topic. Caputo really was aghast at my treatment by the Deep State, and he went to war of sorts for me. I was privy to some very heated arguments he had with senior human resource personnel who were thwarting my every move. I saw him combat the Deep State that worked against us to slow down our work daily to impact the pandemic response.

We got no help, none. Since most of the personnel worked from home, it was easy for them to use this as an excuse. Really, at the height of the pandemic response in mid-2020 onwards, the HHS building was staffed with us only, skeleton staff. Our communications and response unit was the only staff that came to the HHS building routinely, besides the military (army and navy) as part of OWS, the scientists and personnel who worked on the seventh floor for OWS, and those who worked for Moderna.

Of course, the White House personnel (including the White House liaison) were extremely apologetic about what was happening to us as part of the HHS teams (i.e., getting no help from HSS bureaucrats), but appreciated our efforts and my dedication to serve regardless of the obstacles the Deep State had thrown up in my path.

I had the privilege of interacting with Mark Meadows, and always found him to be intelligent and well-informed. Disciplined. I found that his opinion on two critical issues, the catastrophic failure of the CDC's test in the early days of the crisis, as well as his general opinions of the pandemic response, tended to match my own. On the testing issue he wrote:

> Early in the pandemic, for instance, the president's Coronavirus Task Force had let him down. They had relied on the CDC to design and manufacture tests for the virus, which is not something that the agency was capable of. Yet Alexander Azar, the president's Secretary of Health and Human Services, had continued to assure President Trump that everything was fine with tests. Although smart when it came to some things, Secretary Azar was a classic example of someone who tried to please President Trump by never disagreeing with him or saying anything that he believed might upset him. This, as I had learned during my time in Congress, was not the way to go.
>
> Eventually, President Trump discovered the CDC's horrible failings when it came to Covid-19 tests. He learned that the agency could not produce enough tests in the time that was available to them. He also learned that the tests the agency had managed to produce were almost entirely inconsistent and laborious, making them effectively useless.[120]

I felt Meadows was letting Azar off easy for his role in the testing fiasco, since America flew blind regarding the spread of the virus for five weeks due to the devastating testing failure. It's one thing to say that Azar simply didn't want to bring bad news to Trump. It's quite another to allow things to go completely off the rails when the importance to the country is so critical. I don't understand how Azar kept his position when he had so badly handled such an important project. I also questioned, and felt it was wrong to grant liability protection to the vaccine developers and all health agencies, leaving the vaccinated population the only ones at risk. I felt Azar made a catastrophic decision with the PREP Act, which specifically granted immunity from liability for COVID-19 vaccinators. However, it was a pattern I would begin to see all too frequently. For all the ink spilled in the media about Trump being mercurial, I always saw in him a cautious, almost corporate approach. Yes, there was what Trump said that might set some people's hair on fire, but when you looked at the actual policy that was implemented, it was a rational, even-handed one.

With the failure of the CDC's testing regimen, Trump turned to private industry to answer the challenge. As Meadows recalled of the decision:

> It demonstrated a fundamental truth about President Trump—that no matter how difficult a problem may be, it was always best to give him the facts upfront, then work with him to reach a solution. It was also a clear indication that he understood, as a businessman, that the private sector was much better suited to solving our problems than the bloated, inefficient bureaucracy of the federal government.[121]

Mark Meadows had an interesting start in his role as chief of staff, as on March 7, 2020, he was told he had been exposed to somebody with COVID-19 and had to be quarantined at his home for two weeks. But it allowed him time to watch like the rest of the country the rise of the media superstar, Dr. Anthony Fauci.

> On Sunday night, I saw an interview with Dr. Anthony Fauci, who seemed to have designated himself as the nation's primary source for information about the Wuhan Virus. (It would be a few weeks before calling it that was deemed "racist" by the mainstream media.) Normally, I wouldn't have had time to watch the interviews and public appearances that the members of our task force were doing, but the time at home gave me an extra couple of hours every day.
>
> Asked about whether people in the United States should think about wearing masks to stop themselves from catching the virus, Dr. Fauci responded dismissively, using the sage, all-knowing voice we would come to know and despise.
>
> "Right now," he said. "in the United States, people should not be walking around with masks. When you're in the middle of an outbreak, wearing a mask might make people feel a little better and it might even block a droplet, but it's not providing the perfect protection that people think it is."[122]

Yes, the most dangerous place in Washington, DC, was between Dr. Tony Fauci and a television camera. The mental gymnastics Dr. Fauci performed in regard to masks were truly breathtaking to behold, a gold medal performance for the ages.

> When it came to masks, his private message did not match the public one. In private, Dr. Fauci insisted that they were crucial to handling the spread of the

virus. In fact, he said if we didn't get enough masks to hospitals and health-care workers across the country, the effects could be disastrous. At the time, I had no opinion on whether masks were useful in preventing the spread of the virus. Like anything else, I was willing to wait for the evidence and draw conclusions based on data. I eventually asked him if we could avoid a lockdown if we mandated masks at work. His response again contradicted his previous statements. Masks would not protect you enough to go to work. So according to Fauci, masks would work in hospitals, but not in factories.[123]

How had Dr. Fauci ascended to his position at the top of the public health pyramid of the United States, how had he stayed there since 1984, and why was he continuing in that role in 2020 when he made absolutely no sense? It is one of the greatest mysteries of the crisis. The reality is that every single statement on COVID-19 by Dr. Fauci and the COVID response, including the policies and mandates on the COVID gene injections, have been flat wrong, every one of them. Moreover, Dr. Fauci has made statement after statement since February 2020 regarding COVID-19 with no data or evidence to back any of it up.

The questions that Meadows had in March 2020 would be the same ones I would have in May when I began my job. As Meadows wrote in his book:

> Why, I wondered, were we telling people that when they came into contact with a person who might have the virus, they had to quarantine for fourteen days? I had heard this number repeated several times now, mostly by Dr. Robert Redfield and his colleagues at the CDC, but I had no idea where it came from. I had never seen any data to back it up.
>
> Why were we even talking about shutting down schools when every study that had been conducted so far showed that the virus posed virtually no danger to children?
>
> Why were the CDC and the WHO taking the numbers that China was giving us at face value when we knew from our intelligence that those numbers were false?
>
> What was the source for all these numbers that I kept hearing about death rates, infection rates, and the number of people who were probably going to die if we didn't lock down the country immediately? If they were just models based on what was happening in other parts of the world, how were those models being formed, and who was accountable for them? Were there any dissenting voices being heard?[124]

How many times do we have to learn the lesson that it's never a good idea to censor debate, and it's always a good idea to ask for the data upon which you base policies that will affect the lives of hundreds of millions of people? We knew China was lying about the outbreak of the virus and that they were lying about no human-to-human transmission. We knew that WHO was also lying for China when they said no human-to-human transmission had been detected in China of the novel coronavirus, according to the WHO's January 14, 2020 tweet,[125] and we knew China was likely lying about the origins of the virus, since they were a communist dictatorship who valued the lie above the truth, and yet we were accepting their data without question. How could any of this be a good idea?

It wasn't simply that Dr. Fauci was a bad scientist; it's that he was dishonest, duplicitous, and vainglorious. Dr. Fauci became very skilled at obfuscation and misrepresentation of facts. As Meadows recalled:

> To Dr. Fauci, however, data was malleable. So was truth. He said whatever sounded good at the time. Often, he would make things up out of thin air. When someone asked him a question, he would give an answer. It mattered little whether the answer was based on evidence. If the cameras were pointing at him, he was happy.
>
> During one early meeting, I had heard him suggest that eventually, everyone in the United States was probably going to come down with the virus. He said that this was inevitable, but that it would provide us with something he called "herd immunity." Then, just a few minutes later, he was talking about how important it was to make sure everyone stayed inside and off the streets, ensuring they would *never* get the virus.
>
> These two things did not seem to go together from my perspective.[126]

It is not just my opinion, but the opinion of other key players in the administration that Dr. Fauci was a dangerously inept official. In the beginning, in the fog of the COVID war, it may have been understandable that Trump did not want to precipitously fire Dr. Fauci. Meadows further described those early chaotic days:

> But we gave him [Fauci] the benefit of the doubt for a while, chalking those contradictions up to the mercurial nature of pandemics. We assumed that when he said two contradictory things in the same day—or in the same *sentence*, as was often the case—he was doing so based on information we didn't have. By his own efforts, he was profiled in every magazine in the country.

At first, his presence seemed soothing to people, and what President Trump cared about at the time was keeping the American people safe as well as calm.

But it wouldn't be long before we realized what a destructive force this little guy could be —or how much damage he had already done to the country. Over the next few days I watched – or rather, I listened from my home— as Dr. Fauci and a few other "experts" in the room completely dominated the conversation around what steps we needed to take in response to Covid-19. Whenever they spoke, the rest of the people in the room would assume that whatever they were saying was based on hard data or verifiable truth. In most cases, no one even bothered to check their numbers.[127]

I can sense the inevitable question: If this was how credible, powerful individuals in the administration were feeling about Dr. Fauci and others, why didn't Trump simply fire the man and take the political hit? Moreover, not just Dr. Fauci, but Dr. Birx also had to go, yet Trump failed to fire them.

The simple answer is, I don't know. I do not know why he did not do this.

I also try to be sensitive to the immense difficulties in serving in a job such as chief of staff to a president. The chief of staff is the guardian of the president, and that means he's got to take a lot of body blows for his boss. With that being said, I often got the feeling that when my boss, Michael Caputo, called Mark Meadows to talk about some situation, Meadows didn't seem to treat the complaint with the appropriate gravity. I got the sense that Meadows viewed us at HHS and communications as the guys who'd been hired to help with the messaging, and we simply needed to do our jobs without any intervention from the White House. More than once I saw how deeply this wounded Caputo. Often after a conversation with Meadows, Caputo would have this devastated look on his face, and say something like, "They really don't understand how fucked up this situation is." Caputo respected the chief of staff, was deeply committed to the president, and didn't feel the people around Trump had a similar level of commitment. Caputo showed clearly that his loyalty was to Trump, and I respected him for it. He also showed me a deep caring for the good of the nation as a whole.

I also developed my own opinions about Meadows. He was a tough guy and had his own vision of what should be done, and sometimes it didn't seem to align with what Trump wanted. He gave me the impression in his discussions with Caputo that he expected us to align with his guidance, and that even if Trump didn't see things in the same manner, Meadows would eventually get the president to follow along.

Is this a criticism, or simply a description of what often happens to the leader of a powerful organization? Ideally, you surround yourself with other strong individuals, and often in that interchange an even better plan is developed.

However, the COVID-19 crisis wasn't just business as usual. It needed revolutionary action, such as firing people like Dr. Anthony Fauci and Dr. Deborah Birx, and replacing them with outsiders whose loyalty was to the American people, not the organizations that employed them and the industries with whom they had long-standing alliances.

However, what I can tell you is that I and many others urged many in the White House—all the way up to President Trump—to take just such a course of action. The pandemic response was a catastrophic failure in terms of the lockdowns, business closures, and school closures, and we wanted Trump to clean the slate and get the right people in fast. We kept sharing and urging. Dr. Fauci and Dr. Birx had to go.

And I also have to say, as a person who was at the absolute center of the government's response to the COVID-19 crisis, I heard gossipy talk and rumors about what people thought could happen if Trump took such actions; in other words, people would speculate that he would be "taken-out" like President John F. Kennedy was for wanting to wind down our involvement in Vietnam. The actual threat promulgated by these speculations and rumors varied. The views could be as extreme as Trump being directly taken out, and Vice President Mike Pence, a big fan of Dr. Fauci, would take over as president. In other versions of the rumor, one of the Trump kids, Don. Jr., Ivanka, or Eric, would be taken out as a warning. These were speculations, yet it showed me how extreme people's thinking was and how rotten the Washington culture and mindset was. Could Trump have been limited or constrained by threats on him and his family? Is this even possible? Or is this just gossip and rumors? Was Trump hated that much by the bureaucratic Deep State? Did the bureaucratic Deep State feel that threatened by Trump? Or are we really talking about people's minds running wild with too much imagination?

Others claimed it would be a less bloody coup, and more leaks of lies about Trump in the media, as a quote often misattributed to Mark Twain (but more likely descended from a line by satirist Jonathan Swift[128]) that, "A lie can travel halfway around the world before the truth even puts on its shoes."

The bureaucratic Deep State has many ways in which it can enforce its will, short of an obvious takeover.

It is also possible that Trump simply saw these bureaucrats as people he could not fire because it would involve too much controversy. By virtue of the billions of dollars of research money Dr. Fauci had given away over the years, and his penchant for *Godfather* philosophy, Trump might have worried that in one way or another, this diminutive Michael Corleone of public health would find a way to make him pay if Trump fired him.

I cannot vouch for the accuracy of any of these rumors, but they were in the air around Washington, DC, while I was there, in the air around HHS, and in the air around the COVID-19 task force.

In fact, in September 2020, it would all come to a head when my boss, Michael Caputo, the number two man at Health and Human Services, the man most directly responsible for me being hired, and someone I consider to be a serious thinker, would make just such accusations.

***

A decent amount of my time was taken up with Operation Warp Speed (OWS), and as I've said, one of the OWS officers eventually helped me get a salary for the last few weeks I worked and sat in on the task force, so I'm extremely thankful for that. I have huge praise for the US military for the decency it showed me when the bureaucratic Deep State treated me terribly.

I want it to be understood that I am still bound by security and confidentiality oaths I signed when I started working with HHS in support of the task force. I visited the seventh floor, giving my opinions about Operation Warp Speed when it was solicited and even when it was not. I was very abreast of all of the COVD-19 research and science since it was part of my job. I arrived in DC well-versed in COVID given my prior high-level COVID posting and duties with WHO/PAHO. There is much I wish I could tell you about the time I spent in DC, at HHS, and in my work on COVID-19, but it would be illegal for me to reveal much about it.

However, that does not prevent me (and I would say even requires) lending my voice to those who have published research questioning the vaccine, as well as talking about things that I do not consider classified.

I've mentioned before that Operation Warp Speed seemed to be a combined pharma/military operation and, when I would climb the twelve steps from my floor to the seventh floor, I would be facing soldiers with guns. They were armed and in full military gear. Now, I'm as patriotic as the next guy, and of course I quickly became good friends with the guys with guns. I have to be fully candid that I am a huge supporter of the blue, our police, and the military, and these people inspire me by their service and the risks they take.

But this seemed to be another level entirely, as if the military was being deployed in order to protect the pharmaceutical companies; as mentioned, intelligence officials (spooks) also seemed present. However, it didn't seem that the pharma companies were taking direction from the military or the spooks. To me, everybody was taking their orders from the pharma companies.

Indeed, as I previously mentioned, Trump even gave Moderna a head office on the seventh floor, right next to the OWS office. As much as Trump is a maverick, it's still important to realize he's a corporate CEO, and he tried to set a structure, which was going to provide him with information and an action plan.

When people ask my opinion of what went wrong with the COVID-19 response, I say, "Trump trusted the wrong people." Even after everything that has happened, I believe Trump to be a good, possibly even a great man, but one who was deceived on a matter of great consequence to the world. He was greatly misled by Dr. Francis Collins, Dr. Fauci, Dr. Birx, and the pharmaceutical companies Pfizer and Moderna, etc. He was deceived by CDC, NIH, NIAID, and FDA officials. He trusted they were giving him optimal, trustworthy counsel, yet they mislead him, from lockdowns to vaccines.

The unfortunate truth is that those who are at the very top of the power pyramid in our society—whether it be former pharmaceutical executive and the secretary of Health and Human Services under Trump, Alex Azar, or former defense industry lobbyist and Trump's secretary of Defense, Mark Esper—are products of a corrupt system.

I believe Trump looks for people who are winners. It's a small club when you're that rich and successful. Compared to people like Azar and Esper, Trump is a boy scout when it comes to corruption. Trump might certainly have stiffed some subcontractors in building his empire (a common practice usually remedied with some sort of settlement and promises of a better deal on the next job), but he didn't have anything to do with putting out dangerous drugs or salivating over weapons systems in the billions of dollars.

As part of the daily pandemic response, every morning at approximately 7:00 a.m. Eastern, there would be a call involving the top level of all the directors and subdirectors of the various agencies in which we talked about data and the latest news. How does this affect the United States? What messages have to go out? How is the communications department going to respond? I was told by several of my superiors, "Paul, there are career people at the NIH and CDC who will be on these calls. They won't speak, but they're in the call mainly to listen to what we're saying. And depending on what we say, they're going to take it to the press. Their job is to subvert and undercut President Trump and they do this by leaking to damage him. You are going to have to be very careful about what you say. You might think we're having an in-house discussion or trying to figure out what's best for America, and that the calls involve people on the same page, but that's not what's really going on. They're going to leak everything we say. They're going to work with the IT division to get all your communications, and they'll find a way to target you. If the Deep State went as far as trying not to pay you, or keep you from getting your security badges, they're not done with you. If you slip up, they're going to go to the press with something and try to take you down. You have to be very careful. We all have to be very careful."

When I had yearned as a young man to become a Navy fighter pilot like Tom Cruise in *Top Gun*, I figured the enemy would be in front of me, and the entire United States military would be backing me up. Tip of the spear kind of stuff, right? But as a grown man working for the United States government, I was learning that the enemy was all around me. Yes, the call does often come from "inside" the house.

During those times in June, July, and August 2020, I would watch the daily press briefings with Trump, Dr. Fauci, and Dr. Birx, and I could feel the president's frustration coming through the television. Many times at the podium Trump would pause, turn back to the members of the task force, make some plea about opening the schools, stopping the mask hysteria, or letting society get back to normal, only to be met with their stone-cold expressions. He fought them in front of the cameras, and he fought them behind the scenes. I heard it was especially vicious behind the scenes when the teachers' union kept demanding that the schools be closed. Was it because they were genuinely terrified of their students who posed close to a zero-infection risk (statistically, zero risk of becoming severely ill or dying if infected), or that the liberal teachers' union wanted their shock troops of teachers to be in political campaign mode?

* * *

One of the people I was able to interact with for a brief amount of time was Dr. Scott Atlas of Stanford University. Like me, Atlas had been brought in to be a counterweight to Fauci and Birx. His office was in the Eisenhower building, and I had many meetings with him there. He arrived in August 2020, and I got to know him personally and discover he was a beautiful human being, as well as being a technically sound researcher who valued good data, no matter what the consensus of the COVID task force.

As the science writer Michael Crichton once wrote, "The work of science has nothing to do with consensus. Consensus is the business of politics. Science, on the contrary, requires only one investigator who happens to be right, which means that he or she has results that are verifiable by reference to the real world."[129]

Dr. Scott Atlas explained to me that he felt isolated at the White House and that most of the government was working to destroy him. We had long discussions in his office about the business lockdowns, the school closures, and the enormous damage we thought they were doing to the country. He was so very informed with the evidence, and I enjoyed talking with him and sharing as well as learning his thoughts. We talked about the fights with the teachers' unions who wanted to keep schools closed (for no reason that made sense to us other than upsetting parents and making them turn away from President Trump), troubles with the CDC, and our opinion that promising therapeutics like hydroxychloroquine and ivermectin were being misused in clinical trials and prescribed in ways that would never effectively fight the virus. I thought Scott Atlas was one of the smartest people I've ever met in the field of evidence-based research and science. I thought he was a patriot and an unsung hero. I felt he was badly treated by the Deep State bureaucracy and Dr. Fauci and Dr. Birx mainly out of envy. In my view, Scott Atlas was saving Americans while Fauci and Birx were harming them.

The main criticism against Scott Atlas seemed to be that his core training was in radiology. That's true, but you also have to remember that in order to become a radiologist, you have to get your MD. And in addition to his expertise in radiology, Atlas developed the skill to run cost-benefit analyses for public health policies and was an acknowledged expert in the field. This was a skill set that was sorely lacking among government scientists (especially the ones involved in the COVID response), and rather than respect him and capitalize on his expertise, they resented him because he was an existential threat to the power they were attempting to claim during the crisis. His analyses showed that every single restrictive policy the government had implemented was a colossal failure when the costs in health and economic activity were taken into account. That was his key strength, to balance the benefits versus the costs from alternative courses of action.

Dr. Atlas must have put Dr. Fauci and Dr. Birx to shame many times in front of the president by pointing out where they were misreading or misrepresenting the science. I was also told this many times by others who were there. Yet the reality is that even Dr. Atlas did not need to school Trump extensively, for I can tell you Trump did his homework and had discussions that were very in-depth and detailed. He understood and knew the minutiae of the underpinning science even as well as, or better than, those trying to confuse him.

This is what Dr. Atlas wrote of one such encounter with Dr. Birx and the president:

> One memorable pre-briefing during August proceeded with both Birx and me seated in front of the president in an arc of four or five chairs arranged in front of his desk. As usual, close to a half dozen more people were further behind me, seated on the sofas or standing, but I don't remember who was in the room.
>
> To no one's surprise, the president brought up one of his favorite topics—testing. By then, he and I had discussed testing several times—not just the massive testing apparatus that had been developed and not just his frustrations at having "cases" defined by a test, even if the person is not sick. We had gone over the importance of frequent testing in protecting high-risk individuals, its impact on the safety of health care workers, and its pitfalls in keeping schools and businesses open. Testing was a complex policy topic, far more nuanced than the "test, test, test" mantra voiced endlessly by almost everyone in the media. It was a critical tool that had significant value when used properly.
>
> We went through the importance of leveraging the massive testing capacity to save lives. The importance of testing in nursing home staff and the elderly was emphasized. Testing to protect the people who had a significant risk from the virus was the most important point. Testing was very important in all high-risk environments, especially in hospitals and patient care settings. We also discussed how testing should be used to help open society safely, rather than be used to quarantine low-risk, healthy people and shut down the lowest risk environments like schools. Toward the end of this Oval Office pre-briefing, the president asked Birx directly, "Do you agree with Scott on the testing?"

I knew where Birx stood on the issue. Testing had also already been a topic in the Task Force meetings, although that discussion was not fully fleshed out. I looked to my right, where she sat, as she began adjusting her position in the chair. She hesitantly replied, "Yes, I think so," and she looked at me for affirmation. I was emotionless, but her half-hearted statement was a lie. I think the president probably sensed that, so he turned his eyes toward me and asked me, "Scott, is that true?"

Without hesitating, I answered his question. "No, she doesn't agree. Dr. Birx thinks we should be testing healthy, asymptomatic people, and if they are positive, they need to be quarantined for fourteen days. And even if exposed people test negative, they still need to be quarantined." I matter-of-factly went on, eyes straight ahead looking at the president. "And that leads to locking down healthy, low-risk people, and those are the people that make up the workforce. That leads to locking down businesses and closing schools." The president nodded but said nothing.

Since this was the last of many topics covered in this pre-briefing, the president ended the discussion and walked into the room connected to the Oval Office.

We all stood to leave. It felt very tense, but there was zero chance I would lie to the president of the United States. He asked me a direct question, and I answered it truthfully. There was no dilemma, no choice in my mind. Birx apparently felt otherwise. She threw a fit, right there, in front of everyone, as we stood near the door before leaving the Oval Office. She was furious, screaming at me, "NEVER DO THAT AGAIN!! AND IN THE OVAL!!"

I felt pretty bad, because she was so angry. I had absolutely no desire for conflict. But did she actually expect me to lie to the president, just to cover up for her? I responded, "Sorry, but he asked me a question, so I answered it." I glanced at a couple of others in the room who had seen the entire episode and muttered, "Well, that didn't go very well," as I exited.[130]

Unfortunately, I shared Scott's opinion of Dr. Birx. Let me be frank: Dr. Birx detested me, which was made clear in her book. Yet today, every single statement by Scott and I has come to fruition as to the harms of lockdowns, the devastation of school closures, and the ineffectiveness and harms of the COVID face masks, thereby vindicating us. Though Dr. Birx will never ever admit that she and Dr. Fauci were catastrophically wrong on everything, including their love affair with the initial Imperial College modelling that was devastatingly wrong. She, in my opinion, detested anyone who was schooled on the science, came prepared to discuss the issues, and was willing to defend the issue and go to battle. She seemed to be a political creature rather than a scientific one, always weighing how her words might move her preferred path forward, rather than telling the truth and letting the chips fall where they may.

And the fact that Dr. Birx was comfortable lying to the president of the United States staggers my understanding of her reasoning. And when she was caught doing it, she turned her anger on the person who told the truth. Washington, DC, is indeed a crazy place.

In talking about what he observed of Dr. Birx's work as task force coordinator, this is what Dr. Atlas had to say:

> She [Birx] also was key to educating everyone at the COVID Huddles. I had expected complex analyses of special data that others had no access to, but the trends that Birx put forth to the White House communicators, day after day, meeting after meeting, were fundamentally simple tabulations of weekly tallies. Adding to the problem, Birx involved circular reasoning as "proof" that locking down was successful in stopping the spread of cases. Like so many others during the pandemic, she relied on models that predicted a certain number of cases and deaths without any accounting for the cyclical decrease in cases that characteristically occurred as time went on, due to increasing immunity, seasonality, and other factors. Because those continued high levels predicted by her model-of-choice failed to materialize, lo and behold, it must have been due to the success of the interventions in place.[131]

Dr. Birx and Dr. Fauci were intentionally mishandling the pandemic and selling themselves to the American public as the epitome of science, and yet if what they were doing was reviewed by an independent panel of experts (not fearful of losing their portion of Dr. Fauci's billions of dollars of research money), they would quickly denounce this foolishness. Simply put, in my opinion, Dr. Birx and Dr. Fauci met their match in Dr. Scott Atlas and I, including Dr. Peter Navarro, who showed that most, if not all, of what they presented to President Trump was scientific malpractice and specious, highly illogical, and blatantly absurd at times.

* * *

Perhaps it's at least a partial explanation that, as Trump was trying to protect the country from COVID-19, he was also in the middle of a re-election campaign and sought to save other initiatives he felt were just as important to the nation. Trump had to juggle the reduction of tensions with North Korea, the negotiation of better trade deals with China, the border wall, peace plans in the Middle East, and the restoration of the American economy and our manufacturing base.

Trump thought there must be a path to victory, both for himself and the country.

Was it loosening up the lockdowns, therapeutics, or the roll-out of vaccines?

They—Dr. Fauci, Dr. Birx, and the Deep State technocrats at CDC, NIH, FDA, etc.—lied to him about the lockdowns, they lied to him about the therapeutics, and they lied to him about the vaccines. They collectively lied to him daily and conspired to make the United States under him in the pandemic seem ungovernable, unmanageable, chaotic, distraught, and a disaster. This was told bluntly to me by officials in government who I would say are part of the Deep State bureaucracy. The aim was to make the voting public walk away from him. They, Dr. Fauci and Dr. Birx and others, ensured the lockdowns were long, hardened, painful, and destructive so that President Trump would be blamed. Yet it was their lockdowns and school closures that Trump approved. The accrued harms and death from the lockdown policies rests at their feet.

I tried to tell the task force about the poor safety testing of the vaccines, so that there would be a permanent record of what had been done.

Nobody listened.

This leads me, in the next chapter, to tell you about the COVID-19 vaccine lies.

CHAPTER FIVE

The COVID-19 Vaccine Lies

The announcement of a potentially effective COVID-19 vaccine on November 8, 2020 by Pfizer and BioNTech, five days after the election, seemed to promise hope for the end of the COVID-19 crisis.

This is how the news was reported in a press release from Pfizer:

> (Nasdaq: BNTX) Pfizer, Inc and BioNTech SE today announced their mRNA-based vaccine candidate, BNT162b2, against SARS-CoV-2 has demonstrated evidence of efficacy against COVID-19 in participants without prior evidence of SARS-CoV-2 infection, based on the first interim efficacy analysis conducted on November 8, 2020 by an external, independent Data Monitoring Committee (DMC) from the Phase 3 clinical study.
>
> After discussion with the FDA, the companies recently elected to drop the 32-case interim analysis and conduct the first interim analysis at a minimum of 62 cases. Upon the conclusion of those discussions, the evaluable case count reached 94 and the DMC performed its first analysis on all cases. The case split between vaccinated individuals and those who received the placebo indicates a vaccine efficacy rate above 90%, at 7 days after the second dose. This means that protection is achieved 28 days after the initiation of the vaccination, which consists of a 2-dose schedule.
>
> As the study continues, the final vaccine efficacy percentage may vary. The DMC has not reported any serious safety concerns and recommends that the study continue to collect additional safety and efficacy data as planned. The data will be discussed with regulatory authorities worldwide.[132]

Of course, the news coming five days after Trump allegedly lost the presidential election to Joe Biden prompted an angry series of tweets from President Trump, as reported by Jon Cohen in the journal *Science*. On November 11, 2020, Cohen wrote:

> In the wake of the dramatic news of a potentially effective COVID-19 vaccine, President Donald Trump posted a flurry of tweets that claimed its makers, the U.S. Food and Drug Administration (FDA), and Democrats had conspired to suppress the announcement until after the 3 November presidential election. The U.S. company involved, Pfizer, "didn't have the courage to do it before," Trump asserted on 9 November. And FDA and Democrats, he wrote, "didn't want to have me get a Vaccine WIN, prior to the election, so instead it came out five days later."
>
> Initially, Pfizer and its German partner BioNTech suggested they might have preliminary trial results by late October—a timetable Pfizer's CEO, Albert Bourla, projected as recently as 29 September. The timeline was based on a plan that called for an outside panel to take a first look at the efficacy data for the vaccine when a total of 32 cases of COVID-19 had accumulated in the vaccine and placebo groups. But the companies and FDA later agreed on a protocol change that nearly doubled that number and delayed that review.[133]

Do you have the chain of events clearly in your mind? On September 29, 2020, President Trump was told that the results would be available in late October 2020.

That means the announcement of a vaccine would be made before Election Day in November 2020.

Trump's promise to get a vaccine out in record time would be fulfilled.

Then suddenly there's a delay, which just coincidentally pushes it past Election Day.

Trump then supposedly loses the election on November 3, 2020.

Of course, he would be mad.

Wouldn't you?

Cohen continued in his role as press spokesman for Big Pharma:

> There is no evidence, however, that the decision had anything to do with presidential politics. And the companies flat out reject at Trump's claims. "What people believe is their business," Kathrin Jansen, who heads vaccine R&D at Pfizer, told *Science*Insider. "Quite frankly, we had no time and still have no time to deal with politics. We are at this 24/7, thousands of people working

diligently to make this work. And for us, it was never about politics, it was always about just the disaster that we were in the middle of, all of us globally, seeing the devastation and the deaths."[134]

Is it too much to refer to Jon Cohen as a "press spokesman for Big Pharma"? Let's see. Did he perform any investigation of the claim that the announcement of the vaccine was held until after Election Day?

He did not.

All he asked for was a statement from the company that will make billions of dollars from the sale of this product and has no liability for harm caused by side-effects from the vaccine. It's a little like asking the suspect in a murder trial if he's guilty. What do you expect the suspect to say?

You can't credibly accuse Cohen of practicing "evidence-based" journalism.

Maybe Jon Cohen of *Science* should have called me for a comment on the news, as I had been the senior pandemic advisor for the Trump administration, but he did not. He could have even asked Caputo his views.

I would have given him my opinion about the release of the vaccine.

But that would have required Cohen to have the ability to understand complexity.

Because there were two significant lies reported about the COVID-19 vaccine.

And it's important to understand that Trump believed one of the lies.

The first lie is that the US Food and Drug Administration (FDA) and the Democrats did not conspire to delay approval of the Pfizer/BioNTech vaccine until after the November 3, 2020 election. They did.

Trump was exactly right to believe vaccine approval (emergency use authorization [EUA]) was withheld to purposely inflict the maximum amount of damage on his re-election campaign.

The second lie is that the vaccine had undergone adequate safety testing. It had not. However, to this day, Trump clings to the belief that the Pfizer/BioNTech vaccine had undergone adequate safety testing.

I hope that Trump reads this chapter and changes his mind.

Let me tell you the basis I have for making these claims and how I tried to stop it.

* * *

Before I was forced to resign on September 16, 2020, I was concerned with how the FDA was taking short-cuts with safety on their testing of the Pfizer/BioNTech vaccine.

In fact, I was so worried about these short-cuts that I wrote a series of emails and a full length technical paper warning of the dangers of this short-sighted behavior and emailed it to all the members of the COVID-19 task force, as well as the CDC, FDA, etc. I sent it separately and directly to Dr. Hahn, commissioner of the FDA. I wanted there to be a permanent record in government files that their senior pandemic advisor had drawn their attention to the problems and warned against approval of the Pfizer/BioNTech vaccine, either before the election or after it.

Although I have included the entire article later in the chapter, I want you to understand my main concerns. In a typical vaccine trial, there's a control group who does not get the vaccine (they get a comparator, in this case, a placebo) and an experimental group that takes the vaccine.

In order to achieve statistical significance, one has to optimally wait until there are at least five hundred "events" in the control and experimental group (combined) in order to have the slightest idea as to whether the vaccine can be determined to be effective. When the number of events is small (two hundred or less), based on our research at McMaster University, then you are increasing the risk of a biased estimate of effect, which is often an "overestimation" of the treatment effect. Yet the bias can go in either direction: an overestimation or an underestimation of the treatment effect.

In this instance, the "event" we were looking for was the development of COVID-19 infection, as well as any side effects. Ideally the outcomes should have been patient-important such as infection, hospitalization, ICU use, or death.

With the control group, you are getting the normal background noise; that is, what you would generally expect in that large sample group.

Once you've identified the background noise of the control group, you're looking for the "signal" from the experimental group (something different from the control group), either in terms of benefit or side-effects.

As I said, in a trial of this type, it's accepted that you need roughly five hundred events before you can say with any authority that you understand the new intervention is effective and what risks there might be associated with its use. You also want a large enough sample size, you do not want to "stop early for benefit" as you may be stopping at a "random high," and you do not want to give the control group the treatment, as you would effectively be ending the comparative aspect of the study.

However, the FDA stopped the clinical trial after there were 162 mild cases of COVID-19 in the control group and eight cases of COVID-19 in the experimental group who received the vaccine.

They stopped collecting anymore data and declared the vaccine a success. They stopped after 170 total events and closed down any further investigation of either group.

That is methodological insanity and violates every principle of scientific investigation. There were also a reported three thousand subjects who were omitted from the analysis submitted to FDA, and the explanation by Pfizer to FDA was incomplete and lacked explicitness. We still do not know clearly why these three thousand participants were omitted. When we back-calculate them into the analysis, we see that the relative risk reduction (RRR) plunges to about 20 percent. This means the Pfizer-reported 95 percent RRR would not be reached to declare success (50 percent threshold), and, also, the absolute risk reduction (ARR), which is the measure that should have been reported, and not the RRR, was 0.7 percent for the Pfizer vaccine. The number needed to treat or vaccinate (NNT or NNV) would then be 142. The "stopping early for benefit" was a very serious flaw in this study by Pfizer for it did not run to sample size, and by stopping so very early, they were unable to assess the safety of the vaccine. This remained a catastrophic mistake in the COVID trials using the lipid-nano particle mRNA platform (by Pfizer and Moderna), and today we pay the price with myocarditis, myopericarditis, vaccine-induced thrombotic thrombocytopenia, and other severe side effects post-vaccine.

I made sure this was stated by me plainly and repeatedly while at HHS, after leaving HHS, and in my article(s).

I even received a visit at HHS from an official from the FDA asking me if I could retract the paper and my statements on the vaccine research flaws. He told me they wanted me to rewrite the paper because, if it became widely known, it would be a catastrophic failure for the FDA and raise serious questions. I told him I could not because what I said was based on principles of scientific investigation and safety.

A few days later I received a surprising call from another official at the FDA, asking again if I could withdraw the article.

I told her I could not.

The next tact she took was to encourage me to rewrite it in a more neutral manner. "Trump really wants this emergency use authorization," the official told me. "And your article could really mess that up." I was aware that Trump was on the stump, repeating the claim that behind the scenes the FDA had determined the vaccine was safe, but was holding the announcement until after the November 3, 2020 election. I knew Trump was misled since he trusted the vaccine companies and CDC, NIH, FDA, etc. to be guiding him as to the effectiveness and safety of the vaccine. He bought it, that it was safe and effective. "Trump is convinced the vaccine is safe and effective and wants the emergency use authorization to come out before the election," the official insisted.

"Trump is not a scientist," I said. "I am. And I say this vaccine hasn't been properly safety tested and you cannot take a twelve- to fifteen-year process and boil it down to a few months or weeks and deem it is safe, especially for children or pregnant women and the developing child in utero. I am happy to sit down with Trump and explain it to him if I need to."

The FDA official then became a little more philosophical. "Well, regardless of whether you'll change the article, Trump isn't getting the emergency use authorization before the election anyways. Actually, we have moved the goalpost in that we are not interested in safety or efficacy anymore, none of that, we are just interested in putting out an EUA but only 'after' the election. The mission is to ensure Trump gets no credit."

"But you guys have already declared the vaccine is safe. Under your own rules you're supposed to issue the emergency use authorization."

"I just want you to know that we at the FDA will fight you on this article and we're going to make sure the emergency use authorization isn't issued until after the election," she said before hanging up.

What she shared staggered and stunned me. It frightened me. I did not know the FDA would operate this way, so recklessly and dangerously. Trump was right that the FDA was holding up approving the emergency use authorization until after the election.

Trump was wrong that the vaccine was effective and safe, and, while I felt this was a fine needle to try to thread for you in this book, I also felt you should know the work I was doing and the discussions I had. I want you to know how much they worked to damage Trump and subvert his every move. Trump was operating in a world in 2020 (and even before) during the pandemic whereby the very same people at CDC, NIH, NIAID, FDA, within government, his own party, and his detractors in the Democrat party, the Deep State, and legacy media were conspiring against him. Completely. There was nothing he could have done that they would have given him credit for. I so very much admired his strength for the hell he gave them.

I will now provide to you the complete article I submitted to the COVID-19 task force, which they never published or even took heed of, and as a result, which Trump never had the opportunity to read. I wrote this and spread it across government as an employee and even after I departed. I was that bold a person that, if I felt I could defend the technical nature of what I was saying, I should share it, for it is imperative if we have important life-saving information that we share it. Everything I wrote in this paper was directed toward the Pfizer and Moderna COVID gene injections and its scientists and CEOs, for I wanted them all to know that there were people out there who knew of the flaws and that the injections (vaccines) could end up being very harmful. As you see, they are.

* * *

Clinical trials stopped early for efficacy or benefit at risk of overestimating treatment effects and distorting risk-benefit assessments: some guiding principles for the SARS-CoV-2 COVID-19 vaccine search

Paul Alexander[1] MSc, PhD

[1]McMaster University, Evidence-Based-Medicine, Health Research Methods, Evidence and Impact (HEI), Faculty of Health Sciences, McMaster University, Hamilton, Ontario, Canada

Keywords: COVID-19, vaccine, clinical trial, early stopping, interim analysis

What is new?

There is unprecedented need to deliver effective therapeutic and preventative agents in the context of the global COVID-19 pandemic. This commentary emphasizes the methodological consequences of stopping clinical trials early, specifically COVID-19 vaccine trials, and particularly from seemingly beneficial results during multiple, early interim analysis of study data. *This commentary reviews the evidence to declare a needed event number of at least 500 if the vaccine trial is to be stopped early for benefit.* [Bold and italics added by authors.] In addition, it is critical that a prespecified stopping rule explicitly outlines all facets of the stopping rule that includes not too many interim looks at the data by the DSMB. As much sample size accrual is critical and most critically, is the at least 500 events (infections) combined in both trial arms. There must be a stricter stopping rule/boundary of $p<0.001$ if stopping early for benefit is considered.

Commentary

In the context of the quest to produce a vaccine for SARS-CoV-2 coronavirus disease under the present pandemic emergency, it is essential to understand the methodological ramifications in randomized clinical trials (RCTs) that are stopped early for efficacy or benefit (immunogenicity), such trials also known as truncated trials (T-RCTs). Our concern surrounds whether the estimates of effect reflect the true effect of the vaccine. Importantly, we reflect on the safety which can be considered a far more critical consideration than efficacy/effectiveness. We must ensure the latter in this COVID-19 vaccine search with no room for error in this regard.

As a case in point in this rapid race to a COVID-19 vaccine, the United States Operation Warp Speed (OWS) is coordinating ongoing clinical trials development research in tandem with regulatory, manufacturing, and distribution processes to avoid delays in disseminating vaccines to the public to prevent the spread of SARS-CoV-2 coronavirus and potential COVID-19 disease (1). The public-private partnership is innovative and thus far quite favourable in driving towards an efficacious/effective vaccine (s), and involves the synchronous trialing of various vaccine platforms with a multiplicity of vaccines to deliver vaccines. While a near majority of vaccine development is based in the United States, the race for a COVID-19 vaccine is global effort, with China, Asia, Australia, and Europe also centers for rapid vaccine development(2). This commentary guidance applies to all global parties involved in vaccine development and as such, in the process of spearheading vaccine clinical trials.

Applicable to global COVID-19 pandemic vaccine rapid trial approach, are potential risks and justifiable questions on vaccine safety and efficacy, with the demand to ensure that only a safe vaccine will come to market. The public, clinical, and scientific community must be assured that no aspects of a safe and effective vaccine development have been breached before approval of a fast-tracked vaccine. The public must always 'only' receive therapeutic interventions (vaccine) that were based on the highest quality, most robust and trustworthy evidence that is underpinned by high certainty, precise estimates of effect. [Bold and italics added by authors.] This commentary serves as guidance but more as a warning to all COVID-19 vaccine developers, of what must be in place to derive confidence by researchers and the public.

One such area that deserves acute attention surrounds the issue of stopping a trial early for benefit that has been discussed in the media, versus continuation of recruiting trial participants. *No doubt it is best not to stop early when there are initial benefits shown because there is a very high risk of making the decision to stop early based on incomplete and inaccurate trial information. Enough data is likely to have not yet emerged for the estimate of effect to be definitive or as we argue, even accurate. Should we stop the trial if the early benefits are very substantial? Can the observed results possibly be too good to be true? What happens if a trial is not stopped before early indications of efficacy, and then goes on to reveal no effect or even serious harm? Then had the trial stopped early for benefit, this could have been catastrophic for future patients.* [Bold and italics added by authors.] Limited adverse event safety data is a real cause for concern if a trail is stopped early, and researchers must carefully consider and balance this need, and plan to ensure this safety data is clear enough and collected longer-term even when a decision is made to truncate.

Researchers must also consider the substantial ethical elephant in the room of enrolling participants who have a random chance of being assigned to the placebo group when you have earlier indications of a potentially large treatment effect (3). What should be done? Does appropriate adherence to study methodology then exclude control groups from the possibly beneficial treatment? Should we deny the control group a potentially beneficial vaccine? Is it more important to focus on the safety of prospective patients and the larger society so that they do not make treatment decisions based on inaccurate or incomplete or even dangerous information, while there is clinical equipoise? These are the issues that the data safety monitoring board (DSMBs), also known as the data monitoring committee (DMC), who will be assessing vaccine trial data, must confront.

Some RCTs are stopped early when investigators conclude that the magnitude of effects of treatment are so large and not due to random error or chance that they must stop the trial early for benefit and administer the treatment, or vaccine. Our primary concern is that if this is based on an interim analysis of the data, that this could indeed be very misleading and drive inaccurate results. This is also very different to when researchers stop a RCT for futility (4) or harm (5). It is crucial to raise the cardinal issues that must be considered if a SARS-CoV-2 coronavirus vaccine is stopped early for benefit to ensure confidence by the public and the scientific community in the effectiveness and safety of the vaccine candidate.

The following principles are what DSMBs who monitor the safety and efficacy of ongoing RCTs, the trial expert leaders, and the national medication/vaccine safety regulators must consider in their decision making:

1) T-RCTs stopped early for benefit are at serious risk of overestimating beneficial treatment effects and can be very misleading (5-9). *Some health care leaders such as Dr. Anthony Fauci (NIH/NIAID) and Dr. Robert Redfield (Director of the CDC) have gone on record to state that as little as 100 to 150 events (infections) would be needed to know if the vaccine is effective.* [Bold and italics added by authors.] This is inaccurate on its own and certainly a cause for concern if this is the threshold being considered. Hughes and Pocock (7) were prescient in their clarion call on the surprisingly broad and imprecise, misleading observed treatment effects that emerge in clinical trials that stop early, and a close examination of the published evidence do indicate that T-RCTs are routinely associated with greater effect sizes and overestimation than RCTs that did not stop early(10).

What do we mean by this overestimation or inflation of the estimate of effect? The overestimation is due to random error or chance since T-RCTs can yield results that are located at the high end of the random distribution of results(11). This is referred to as a 'random high'. More considerable and random differences from the true treatment effect can emerge early on in the trial when the sample size is small, or the number of events is smaller, these two a result of stopping early before the trial can run to its powered (based on primary outcome) sample size. If the sample size is small or the number of outcome events is negligible when the data safety monitoring board (DSMB) takes an early look at the data, then the effect estimate will need to be of greater magnitude to meet the standard, prespecified stopping rule boundaries to control Type I error rates (avoidance of a false positive) if interim analyses are planned(4). DSMBs overseeing vaccine trials for COVID-19 must be mindful of the sample size and number of events as they assess efficacy, as well as the stopping rule used when they take interim looks at the accumulating data.

2) The prespecified stopping rule is a critical consideration for the DSMBs. If the data is checked periodically and investigators make a decision to stop the trial as soon as a large magnitude of effect is observed, then this could lead to overestimation of the effect of treatment(11). The problem arises with the repeated interim analysis of the data with no formal process, as opposed to formalized *a priori* rules in the interim analysis process. Moreover, *a priori* defined stopping criteria or a more strict boundary of p<0.001 will also work when having prespecified interim analysis, to mitigate early stopping(11). Caution is urged in setting the conditions for early stopping, and it is critical to report openly and transparently if there were formal rules defined *a priori* before a study is indeed stopped early. Researchers must be explicit

in this stopping rule. Experts agree that routine data monitoring practice demands a predefined statistical stopping rule (10).

Evidence suggests that small trials with very large effects may be due to early stopping due to repeated interim analysis of the data. Regular looks at very short intervals, e.g., every 5 patients, is very problematic, as there are likely wide swings of the point estimate. An *a priori* formal rule with fewer, more infrequent interim analysis (and potential adjustments for the multiple analyses), larger intervals between each analysis, a strict stopping boundary (p< 0.001), and large sample size will work to reduce the chance of overestimation (11, 12). At the same time, some argue that a strict stopping boundary based on p-value may increase the risk of an inflated estimate even as it guards against early stopping. Thus, to overcome this, the call is for lesser looks at the data and looks that occur much later in the trial when a sufficiently larger number of events and sample size has accrued (11, 12). All of these aspects can increase certainty (confidence) in the estimate of effect and all things considered, it is the number of events that really drives this confidence in the estimates of effect.

3) We focus on what we consider to be a key rate-limiting step in stopping early for benefit and the troubling risk of overestimating the effect and declaring benefit when there is none. Increased events will reduce the likelihood of overestimation of effect (10). The goal is to ensure that the treatment effect is as accurate as possible and events play a core role in this assessment. Experts agree that a core aspect of any prespecified stopping rule must include a sufficiently large number of outcome events. *As events accrue (for COVID-19 vaccine research, 'events' are the infections), the risk of an inflated estimate is reduced. Credible research suggests that this number appears to be at least 500 outcome events (10) for the estimate of effect to be more likely near the truth (10). Five hundred events as a threshold are very different from the 100 to 150 or so events (infections) publicly alluded to by Dr. A Fauci and Dr. R Redfield. This raises a serious question on the optimal minimal number of events to declare efficacy/effectiveness in COVID-19 clinical trial research.* [Bold and italics by authors.] As an example, a systematic review and meta-analysis (employing multilevel meta-regression) by the globe's top research methodologists looking at T-RCTs versus RCTs that were not stopped for the same research question, found that there were very large overestimates of effect when the trial was stopped early and had outcome event numbers less than 100, large overestimates when event number was 200, and still appreciable overestimates of effect when event numbers

were between 200-500(10). Trials with smaller sample-sizes, and small events stopped early are a serious issue as to spurious and inaccurate results.

This weakness or fragility in the estimates of effect appears the same even when the trial has a larger sample size if the number of events is small. Sample size has less to do with the accuracy of the results and in the case of COVID-19 vaccine research, the reported large trials with 30,000 sample size has less to do with the estimate of effect than the number of events. All this has led researchers to warn of skepticism when *any* trial is stopped early for benefit and mainly when the outcome event number is low (<500)(10, 11). Simulations and recent empirical evidence do reveal that T-RCTs can miscalculate and lead to large, misleading, overestimated treatment effects when there is a small number of events (<200)(13). Therefore, DSMBs associated with COVID-19 vaccine trials such as OWS or any clinical trial globally, should be very mindful of the minimal number of events that would drive confidence in the estimates before the decision to truncate. There will be far less confidence in the vaccine trial's estimates of effect (result) if stopped early for benefit and the number of accrued events (infections) is less than 500.

4) The most suitable choice of primary outcome/endpoint and stopping guidelines is also a vital decision consideration for the DSMB and trial leadership, and the clinical relevance should drive the primary endpoint (and secondary) decisions. Patients need information from patient-important outcomes so that they are adequately informed for their values and preferences decision making (8). The outcome information must be clear so that patients can fully weigh the benefits as well as the harms of any intervention or vaccine. This must be set a priori and not be trifled with so as to declare success. Key consideration and balancing is also needed for what is the likely sample sizes, expected event rates, and intended duration of the trial (14).

5) **With a specific focus on adverse events, if a vaccine trial is stopped early for benefit, the trial sponsors and regulators must ensure that a phase IV, post-decision pharmacovigilance study is established to follow the patients longer-term to collect data on emerging adverse events. This is critical. It is also critical to ensure confidence and safety post-marketing by providing that such longer-term surveillance (safety trials) is established to monitor safety throughout the vaccine's life cycle.** [Bold and italics added by authors.] Having adverse events emerge before stopping for benefit or before any assessment is critical to inform whether to continue the trial or stop it for harms. This was just the case in the recent stoppage of the AstraZeneca (in collaboration with University of Oxford) COVID-19 vaccine trial for a suspected serious adverse reaction (15). It is not atypical for trials

to be paused to assess how serious the adverse reaction was. This is especially important when a study is stopped before evidence of potential adverse events have sufficiently accumulated. This is also a critical issue with concomitant vaccination with a new vaccine in the environment of other recommended vaccinations, e.g. the seasonal influenza vaccine etc. It is thus critical for further follow-up to ensure the safety is clarified into the longer term.

Understandably, a decision to stop early for benefit is fraught with statistical considerations, ethical issues, and practical issues, and these are the very issues that the COVID-19 vaccine experts must grapple with as they search for a safe and effective vaccine. Moreover, a significant concern is the chilling effect (freezing effect) T-RCTs have on future research (6, 13) and particularly as mentioned, the limitation in assessing the potential harmful events that could have accumulated had the trial run to its powered sample size. If a T-RCT has prespecified stopping rules and guidelines with infrequent interim analysis of the data conducted later in the study when more sample size and events have accumulated, and have met the risk of bias criteria (optimally judged at low risk of biased estimates) with the number of events being at least 500 or greater, and with a low p value as a stopping boundary, then one can be more confident that the estimate of effect reflects the true treatment effect. *At least 500 events will allow much more precision and the patient's values and preferences (16, 17) would be more assured in that they are considering a treatment (or vaccine) based on more confidence in the estimates of effect. Bassler et al. (2008) insists on continuation of enrollment and follow-up for a longer period (9).* [Bold and italics added by authors.]

The DSMB of a vaccine trial will also have very critical decisions to make as vaccine research continues and particularly the decision to stop early if a benefit is indicated. Importantly, never before has the medical regulatory agencies been tasked with a future decision that will impact hundreds of millions if not billions of lives and potentially steer public-health policy decision-making globally in terms of an effective vaccine for COVID-19. In this, the DSMBs has a vital and ethical role in ensuring that any beneficial intervention is given to all patients once there is an indication of efficacy/effectiveness. However, as indicated, confidence in the evidence and decision must be based on a consideration of the benefits and the risks at all instances of interim looks at the data, and as discussed, several issues must be considered and balanced typically simultaneously in deciding to stop early. This issue of balancing the benefits versus risks that have accumulated at each interim look at the data is critical and complex and deserves somber reflection and DSMBs and national drug/medications regulator, must weigh this very carefully.

Additional considerations in stopping early for benefit, is the trial would entail a high risk of bias and there will be less confidence in the purported estimates of effect of potential vaccine. Indeed, under the *Grading of Recommendations, Assessment, Development and Evaluation* (GRADE) framework, treatment effect estimates in outcomes associated with studies rated as high risk of bias would lead to rating down the level of certainty. Lowered certainty of a COVID-19 vaccine intervention in a prevention outcome under the GRADE framework, would imply less confidence in the treatment effect estimates, and reduced certainty that further research would not change the treatment

Rapid COVID-19 accelerated vaccine development programs in terms of the clinical development, the process development, the manufacturing, and distribution scale-up has been significantly accelerated in innovative programs such as OWS. The US and world need a safe and effective vaccine. This rapid innovation to address this pandemic emergency should be given the recognition and credit it deserves given the risk and the synchronous coordination of many processes to achieve success. The global scientific community must also be applauded for the rapid scale and immense amount of research that has transpired in response to COVID-19. In closing, this commentary is a caution that the step to stop early for benefit and stop randomization of patients to the potential of no treatment in these COVID-19 vaccine trials should not to be taken lightly. ***Safety must be the core consideration as we strive for a vaccine. We are trusting in the sound judgement, scientific depth, and integrity of regulators and associated experts and at no time in COVID-19 vaccine development, must safety be in question.*** [Bold and italics added by authors.]

Outlined here are the key points to consider (Box 1) in COVID-19 vaccine development. These points (Box 1) once part of these vaccine trials, will significantly allow for increased confidence in the resulting estimates of effect and decisions. Failure to consider these will significantly reduce confidence by the research community and the public in the results. Openness, transparency, and explicitness must be the watchword and the public and global scientific community must be allowed the background data to the extent possible with a consideration of even more openness given this emergency and justified questions. This will allow researchers and the public to be able to understand the decision-making that was involved in stopping early for benefit, any emergency use authorization (EUA), or adverse events that emerged. The public must always be informed by full, comprehensive, balanced reporting and decisions that match the underlying data. Informed buy-in and decision-making by the public is critical and they can only do this with full openness, transparency, and explicitness by the COVID-19 vaccine developers. Once again, assurances of safety is the critical rate-limiting step in this vaccine, and in this, using the optimal 'minimal' number of events of 500 as a threshold to stop early for benefit. We caution to be very careful if any COVID-19 trials are stopped early for benefit and to closely adhere to the guidance outlined in this commentary.

Box 1: Key points to consider in COVID-19 clinical trials stopped early for benefit

---

Key points to consider:

1. Need for a prespecified stopping rule is critical that explicitly outlines all facets
2. Not too many interim assessments (looks) at the data is critical
3. Adjustments for the multiple looks at the data
4. Interim looks late in the trial to allow for larger sample accrual size is optimal
5. Interim looks late in the trial allowing for larger number of event accrual is optimal
6. A need for a stricter stopping rule/boundary of $p<0.001$
7. Event number at least 500 is optimal to protect for overestimation of estimate of effect
8. Stopping early limits accruing of important adverse events data downstream
9. Pharmaco-vigilance surveillance for adverse effects long term is critical
10. Risk-benefit assessment at early interim looks precludes an accurate assessment of risk
11. Continuing the trial to the powered sample size is optimal (follow-up for a further period)
12. Consideration of the patient-important primary outcome (s)

Roles in the study and manuscript:
P Alexander: Conceptualization, writing, final manuscript.

Declaration of interest:

The author holds expertise in evidence-based medicine and guideline development. The author is a part of the GRADE Working Group.

Funding acquisition: There was no funding for this research topic.

References

1. Slaoui M, Hepburn M. Developing Safe and Effective Covid Vaccines - Operation Warp Speed's Strategy and Approach. *N Engl J Med* 2020.

2. Le TT, Cramer JP, Chen R, Mayhew S. Evolution of the COVID-19 vaccine development landscape. *Nat Rev Drug Discov* 2020; 19: 305-306.

3. Prutsky GJ, Domecq JP, Erwin PJ, Briel M, Montori VM, Akl EA, Meerpohl JJ, Bassler D, Schandelmaier S, Walter SD, Zhou Q, Coello PA, Moja L, Walter M, Thorlund K, Glasziou P, Kunz R, Ferreira-Gonzalez I, Busse J, Sun X, Kristiansen A, Kasenda B, Qasim-Agha O, Pagano G, Pardo-Hernandez H, Urrutia G, Murad MH, Guyatt G. Initiation and continuation of randomized trials after the publication of a trial stopped early for benefit asking the same study question: STOPIT-3 study design. *Trials* 2013; 14: 335.

4. Jitlal M, Khan I, Lee SM, Hackshaw A. Stopping clinical trials early for futility: retrospective analysis of several randomised clinical studies. *British journal of cancer* 2012; 107: 910-917.

5. Walter SD, Guyatt GH, Bassler D, Briel M, Ramsay T, Han HD. Randomised trials with provision for early stopping for benefit (or harm): The impact on the estimated treatment effect. *Statistics in medicine* 2019; 38: 2524-2543.

6. Wang H, Rosner GL, Goodman SN. Quantifying over-estimation in early stopped clinical trials and the "freezing effect" on subsequent research. *Clin Trials* 2016; 13: 621-631.

7. Hughes MD, Pocock SJ. Stopping rules and estimation problems in clinical trials. *Statistics in medicine* 1988; 7: 1231-1242.

8. Pocock S, White I. Trials stopped early: too good to be true? *The Lancet* 1999; 353: 943-944.

9. Bassler D, Montori VM, Briel M, Glasziou P, Guyatt G. Early stopping of randomized clinical trials for overt efficacy is problematic. *J Clin Epidemiol* 2008; 61: 241-246.

10. Bassler D, Briel M, Montori VM, Lane M, Glasziou P, Zhou Q, Heels-Ansdell D, Walter SD, Guyatt GH, Group S-S, Flynn DN, Elamin MB, Murad MH, Abu Elnour NO, Lampropulos JF, Sood A, Mullan RJ, Erwin PJ, Bankhead CR, Perera R, Ruiz Culebro C, You JJ, Mulla SM, Kaur J, Nerenberg KA, Schunemann H, Cook DJ, Lutz K, Ribic CM, Vale N, Malaga G, Akl EA, Ferreira-Gonzalez I, Alonso-Coello P, Urrutia G, Kunz R, Bucher HC, Nordmann AJ, Raatz H, da Silva SA, Tuche F, Strahm B, Djulbegovic B, Adhikari NK, Mills EJ, Gwadry-Sridhar F, Kirpalani H, Soares HP, Karanicolas PJ, Burns KE, Vandvik PO, Coto-Yglesias F, Chrispim PP, Ramsay T. Stopping randomized trials early for benefit and estimation of treatment effects: systematic review and meta-regression analysis. *JAMA* 2010; 303: 1180-1187.

11. Guyatt G RD, O Meade M, and CooK D. . Users' guides to the medical literature.: JAMA Evidence; 2015.

12. Pocock SJ. When (not) to stop a clinical trial for benefit. *JAMA* 2005; 294: 2228-2230.

13. Bassler D, Montori VM, Briel M, Glasziou P, Walter SD, Ramsay T, Guyatt G. Reflections on meta-analyses involving trials stopped early for benefit: is there a problem and if so, what is it? *Statistical methods in medical research* 2013; 22: 159-168.

14. Zannad F, Gattis Stough W, McMurray JJ, Remme WJ, Pitt B, Borer JS, Geller NL, Pocock SJ. When to stop a clinical trial early for benefit: lessons learned and future approaches. *Circ Heart Fail* 2012; 5: 294-302.

15. Beasley D. AstraZeneca suspends leading COVID-19 vaccine trials after a participant's illness. Reuters; 2020.

16. Zhang Y, Coello PA, Brozek J, Wiercioch W, Etxeandia-Ikobaltzeta I, Akl EA, Meerpohl JJ, Alhazzani W, Carrasco-Labra A, Morgan RL, Mustafa RA, Riva JJ, Moore A, Yepes-Nunez JJ, Cuello-Garcia C, AlRayees Z, Manja V, Falavigna M, Neumann I, Brignardello-Petersen R, Santesso N, Rochwerg B, Darzi A, Rojas MX, Adi Y, Bollig C, Waziry R, Schunemann HJ. Using patient values and preferences to inform the importance of health outcomes in practice guideline development following the GRADE approach. *Health Qual Life Outcomes* 2017; 15: 52.

17. Munoz-Velandia O, Guyatt G, Devji T, Zhang Y, Li SA, Alexander PE, Henao D, Gomez AM, Ruiz-Morales A. Patient

Values and Preferences Regarding Continuous Subcutaneous
Insulin Infusion and Artificial Pancreas in Adults with Type 1
Diabetes: A Systematic Review of Quantitative and Qualitative
Data. *Diabetes Technol Ther* 2019; 21: 183-200.

* * *

I hope you'll forgive me for dumping my entire article about the lack of
safety testing for the COVID-19 vaccines on you, as well as the challenges I
saw and the issues I felt that the FDA (and CDC and relevant scientists) had
to tend to, but I think it's important for the historical record.

The lowest optimal threshold for determining the effectiveness and
safety of a COVID-19 vaccine should have been at least five hundred infections between the control and experimental groups (and, based on our prior
research, at the least two hundred, so that we could be confident that the
estimate of effect was not biased or overestimated).

They stopped after 170 infections.

This was the biggest gamble in history with the health of the public, and
they were cutting corners with safety testing. It is similar to the catastrophic
and unscientific present EUA approval that the FDA granted to Pfizer
(August–September 2022) for the new, updated bivalent vaccine (Wuhan
legacy strain and BA.4/BA.5 sub-variants) booster (fifth injection) based on
rodent mice data, and specifically using eight to ten mice (not human trial
data) that were examined for a few weeks.[135] The troubling aspect of this was
that the mice still got the virus and infections in the lungs and nostrils, and
the only human data was for the BA.1 sub-variant, which is not predom-
inant and not relevant to the existing BA.4/BA.5 sub-variant clades. My
argument is that you cannot extrapolate the BA.1 human data to the BA.4/
BA.5 mice data, which is what FDA has done. The bottom line is the FDA
approved the EUA to vaccinate 200 million people based on eight mice. Just
like how the FDA gave EUA for the legacy Wuhan vaccine from Pfizer (and
Moderna) based on 170 events, to then go about vaccinating hundreds of
millions of people. Does any of this make any sense? In my mind, the FDA
has completely failed the American people.

In addition to stopping the study after 170 infections (events), they also
abandoned any plans for continued monitoring of the health of the two
groups.

Let me say that again so it really sinks in. We have COVID gene injections (vaccines) approved by the FDA, which have been given to hundreds of millions, if not billions of people in America and around the globe based on no proper safety testing. None! Moreover, all of the COVID vaccine trials to date have been run for weeks and months, and not for the proper years of follow-up duration needed so as to "exclude" harms.

They abandoned any plans for continued monitoring of the health outcomes of the two groups, and, by giving the placebo group the vaccine (unblinding), they effectively ended the study. When we are told that the vaccine trials are ongoing, both Pfizer and the FDA know that this is a lie to the public. There can no longer be any comparative effectiveness data as both trial arms got the same intervention.

It's like conducting a study on the health risks of smoking, then closing the study after two months and saying the health outcomes of the two groups with regard to lung cancer were exactly the same.

What has been the result of data collected on COVID-19 vaccine adverse events?

As of August 26, 2022, there have been 1,394,703 reports of adverse events from the COVID-19 vaccine made to the government's own CDC Vaccine Adverse Events Reporting System (VAERS), which also includes nondomestic reports.[136] I think it's also important to note that VAERS is a voluntary reporting system, which means that physicians are not legally required to submit a report if their patient suffers a reaction, and there are no legal consequences for failing to make such a report. This has led to allegations that, if anything, these numbers represent a massive underreporting of suspected adverse reactions (estimates are that VAERS collects only from 1 percent to 10 percent of the burden). I will go into that in more depth later, but for now, here are the latest numbers as of early September 2022.

- 30,605 reported deaths[137]
- 175,020 reported hospitalizations[138]
- 134,530 reports of recipients needing to go to an urgent care facility[139]
- 204,343 reported doctor visits[140]
- 9,979 reported cases of anaphylaxis[141]
- 15,945 reported cases of Bell's Palsy[142]
- 4,992 reported miscarriages[143]
- 16,385 reported heart attacks[144]
- 51,879 reported cases of myocarditis/pericarditis[145]

- 57,310 reported cases of permanent disability[146]
- 8,942 reported cases of thrombocytopenia/low platelet count[147]
- 33,832 reports of life-threatening medical emergencies[148]
- 44,576 reported cases of severe allergic reactions[149]
- 14,587 reported cases of shingles[150]

Probably the most concerning to the average person is the 30,605 reported deaths following COVID-19 vaccination.

I find it interesting to compare the reported death rate after vaccination across the United States in the five years prior to the COVID-19 vaccination program with the information we have now on just a little under two years of data.

- In 2016, there were 437 reported deaths following a vaccination.[151]
- In 2017, there were 467 reported deaths following a vaccination.[152]
- In 2018, there were 535 reported deaths following a vaccination.[153]
- In 2019, there were 602 reported deaths following a vaccination.[154]
- In 2020, there were 420 reported deaths following a vaccination.[155]
- In 2021, there were 21,884 reported deaths following a vaccination.[156]
- In 2022, there were 9,598 reported deaths following a vaccination.[157]

The 2020–2021 timeframe shows a more than fifty-two-fold increase in reported deaths to VAERS following a vaccination. In science, we generally refer to such data as a signal, flag, or "clue." It may not be the definitive smoking-gun evidence required to silence all the "science-deniers" out there, but certainly something that demands further investigation.

What percentage of vaccine reactions are reported to VAERS? If we consider vaccines to be just like any other medical intervention, the number is likely to be quite low. In 1993, the former head of the FDA, David Kessler, estimated that only about 1 percent of serious adverse events get reported.

> A recent review article found that between 3% and 11% of hospital admissions could be attributed to adverse drug reactions. Only about 1% of serious adverse events are reported to the FDA, according to one study.
>
> There are probably several reasons why some serious events are not reported to either the FDA or the manufacturer. First, when confronted with an unexpected outcome of treatment, physicians may not consider drug-induced or device-induced disease, but rather consider the event to be related to the course of the disease.

Unfortunately, this may be due to the limited training medical students receive in clinical pharmacology and therapeutics. A 1985 survey of US medical schools found that only 14% of them had required core skills and principles of therapeutic decision making and clinical pharmacology. Of the remainder, 87% taught only a few hours of clinical pharmacology, and most of the teaching occurred in the early years of medical training.[158]

Did you have the slightest idea that somewhere between 3 percent and 11 percent of all hospital admissions were likely due to adverse drug or medical device reactions? And I'm sure we'd all like to know whether that number is closer to 3 percent or 11 percent. How is it that we don't have some clarity on these numbers?

Where is the robust surveillance system that should catch these reactions?

The truth is that such systems do not exist. The FDA should have demanded that the vaccine developers (all of them) conduct post-market surveillance of adverse reactions, to investigate each death (including with autopsies and full reports), and to implement data safety monitoring boards and critical event committees, etc. Again, critical to the American people is a proper COVID-19 vaccine surveillance system across all fifty states so that we can monitor the outcome of vaccination, as well as offer treatment and supportive care to all those who are vaccine injured. None of that, nothing, has been done by the federal government of the United States, not by the CDC, the NIH, or the FDA. No one. This should have been mandated by the FDA to Moderna, to Pfizer. Yet the FDA has not done this. Why? That we (Azar at the HHS helm) could grant, via the PREP Act, liability protection to all of the players yet also not ensure we, at the least, have surveillance to properly monitor what is happening post-shot? This is outrageous. This is, in my opinion, near criminal.

In October 2021, a professor from Columbia University, Spiro Pantazatos, and a researcher from Israel, Herve Seligmann, investigated what the genuine rate of COVID-19 vaccine-induced side-effects might be. As they detailed the problem:

Accurate estimates of COVID vaccine-induced severe adverse event and death rates are critical for risk-benefit ratio analyses of vaccination and boosters against SARS-CoV-2 coronavirus in different age groups. However, existing surveillance studies are not designed to reliably estimate life-threatening event or vaccine-induced mortality risk (VMR). Here, regional variation in vaccination rates was used to predict all-cause mortality and non-COVID deaths in subsequent time periods using two independent, publicly available datasets from the US and Europe (month and week-level resolutions, respectively).[159]

Can it genuinely be true that we have this massive research investment in vaccines, an unprecedented marketing campaign involving both corporate and governmental entities, and we don't have a reliable surveillance system?

And what did the two researchers estimate to be the difference between the death numbers reported through VAERS as being vaccine-related, and what their estimates showed?

> The US CDC data allowed for estimation of VMR [vaccine mortality risk] and vaccine-induced deaths. Importantly, our calculations do not rely on VAERS and its associated limitations. Our estimated national average VMR of 0.04% is 20-fold greater than the CDC reported VMR of 0.002%, suggesting vaccine associated deaths are underreported by at least a factor of 20 in VAERS. The estimate is based only on significant effects detected in our analysis, and hence likely represents a lower bound on the actual reporting factor.
>
> Interestingly, our estimates of 133K to 187K vaccine-related deaths are very similar to recent, independent estimates based off of US VAERS data by Rose and Crawford. The authors report a range of estimates depending on different credible assumptions about the VAERS underreporting factor and percentages of deaths definitely caused by vaccination based on pathologists' autopsy findings . . . This factor, multiplied by the number of reported VAERS deaths and the percentage of VAERS deaths believed to be caused by vaccination, based on pathologists' estimates, yields various estimates with an average around 180K deaths.[160]

This research, published in October 2021, less than a year after the approval of the Pfizer/BioNTech vaccine, suggests that somewhere between 133,000 and 187,000 Americans died from vaccine side effects. However, if we accept that only a portion of the adverse effects or deaths are reported and if we say it is the upper limit, so 10 percent, then we are really looking at about one million deaths linked to the COVID vaccine (gene injection).

Multiple researchers are finding that only about 5 percent of suspected vaccine reactions are being reported to the VAERS system, meaning we have to multiply those numbers by twenty in order to get an estimate that reflects reality. If so, then we are looking at over two million deaths.

That means that vaccine side effects, currently listed at 1,394,703 events, should be reinterpreted as likely being 27, 894,060 events.

That means the number of COVID-19 vaccine deaths, currently reported at 30,605, is probably closer to 612,100. These numbers are stunning, and it is difficult to wrap our heads around this.

I believe the COVID-19 vaccine will go down in history as the worst public health disaster of all time, challenged only by the Fauci and Birx lockdown lunacy policy.

CHAPTER SIX

The Destruction Derby (and One More COVID-19 Story)

When are people most likely to tell the truth?

I believe it's when you have tried your best, but it's clear all your efforts have failed to change the situation. For Michael Caputo, the number two official at Health and Human Services, the administration's top representative and spokesperson for the COVID-19 crisis, that moment came in September 2020. This purge would also drag me along with it. This is from a September 14, 2020 article in the *Washington Post*, setting the stage for its readers of the executions to follow:

> Trump installed Caputo in April after weighing whether to fire Health and Human Services Secretary Alex Azar over a series of damaging stories about Trump's handling of the pandemic, according to three current and former White House officials who spoke on the condition of anonymity to describe behind-the-scenes discussions. Allies persuaded Trump not to make such a change amid a pandemic, but instead to being in Caputo, the officials said. (Trump denied reports that he was considering firing Azar at the time.)[161]

The media isn't so difficult to understand. You just need to translate their lies into something resembling truth.

Azar was leaking to the press in ways designed to hurt President Trump and got caught. In my mind, Azar was disloyal to President Trump and thus had been caught, just like I thought.

The only question was what to do with Azar: fire him or try to clip his wings by bringing in somebody like Caputo?

And what about the "three current and former White House officials who spoke on the condition of anonymity"? Is it one current official and two former White House officials, or is it the other way around? How can we assess the credibility of these sources? Are they allies of Azar, doing his bidding, like all the little Dr. Fauci minions sprinkled throughout government? What was clear to me was that Washington, DC, operated only on leaks and scandal, and the daily push on both sides, Republican or Democrat, was to see who could leak first and what the follow-up leaks would be. It really is a putrid, hot mess swamp of a place to live and work, if you are within the government. The entire system operates on subversion, leaks, lies, spin, and a corrupted media.

When you look at the news through my lens, it's all so clear.

Leaking, as mentioned, is a way of life in Washington, DC, and nobody ever gets punished for it.

How do I read that paragraph from the *Post*?

Azar was a leaker and got caught. Trump wanted him fired. But some "allies" around Trump convinced him not to fire Azar, but to load up the government with even more people and make the work of the task force appear more chaotic.

What do you call an "ally" who keeps giving you bad advice?

I'd say you've got somebody who's working for the other team.

I always felt like President Trump was surrounded by people who didn't have his best interests at heart, or that of the country. He was surrounded by people who, with each move, tried to damage him. And even President Trump couldn't see it all the time. Yet the task force was undermining him in conjunction with the Deep State bureaucracy, the CDC, NIH, FDA, and NIAID officials, at each move. They were determined to make President Trump's pandemic response a failure, yet it was their policies he was approving. It is and was the most remarkable feat to witness. Sometimes President Trump caught a glimpse of it and sent out a nasty tweet or two, but then he'd fall back into believing they just had different opinions, and that diversity was a good thing. The *Washington Post* article continued:

> Almost immediately, Caputo began exerting control over officials' public appearances and statements: by early summer, he had extended that scrutiny to scientists. He and an adviser [Author's note: That's me!] have faced mounting criticism in recent days for interfering with the work of scientists at

the Centers for Disease Control and Prevention, seeking to change, delay or kill weekly scientific reports that they thought undermined Trump's message that the pandemic is under control. Caputo has also sought to wield influence over when government scientists appear on television, telling officials that he approves such bookings.[162]

It's easy to understand the media if you simply say to yourself: everybody they want you to hate is probably a good guy, and everybody they want you to love is probably a lobbyist for an industry that supports the media or will be in that position when their government job is over.

You might shake your head and say that's too cynical a view. Maybe that's the way things happened in the past, but not now. If you think our government got cleaned up, let me know when that happened. It didn't. The bad guys just got better publicists and learned a few psychological tricks to fool the public.

From April 2020 until September, I watched Michael Caputo put everything on the field. I was surprised by his love of country and his deep loyalty to President Trump, when most people around him were duplicitous, corrupted, and seeking their own interests. Most around President Trump were not loyal and obstructed his agenda at every turn. In my opinion, they were mini versions of the prior speaker of the house, Paul Ryan, who impeded and subverted President Trump's agenda. Remember, Caputo, in 1994, had moved to Russia and was instrumental in helping President Boris Yeltsin win a second term over the communist old guard of the Soviet Union. He didn't scare easily and had excellent political instincts. I would tell you, exceptional! But by September 2020, Caputo was exhausted, had some medical problems, and poured out his frustrations in a twenty-six-minute-long video on Facebook, which was discussed in a *New York Times* article:

> To a certain extent, Mr. Caputo's comments in a video he hosted live on his personal Facebook page were simply an amplified version of remarks the president himself has made. Both men have singled out government scientists and health officials as disloyal, suggested that the election will not be fairly decided, and insinuated that left-wing groups are secretly plotting to incite violence across the United States.
>
> But Mr. Caputo's attacks were more direct, and they came from one of the officials most responsible for shaping communications around the coronavirus.

C.D.C. scientists "haven't gotten out of their sweatpants except for meetings at coffee shops" to plot "how they're going to attack Donald Trump next," Mr. Caputo said. "There are scientists who work for this government who do not want America to get well, until after Joe Biden is president."[163]

If the media had the slightest bit of curiosity, you would think they would have been interested in understanding how the individual most intimately involved with communicating to the public about the COVID-19 crisis had come to hold such a dark vision of what lay ahead. But they didn't. Caputo gave a detailed account of the future, which he believed lay ahead. It is interesting, in 2022, to review his predictions and compare it with what actually happened in 2020.

Mr. Caputo predicted that the president would win re-election in November, but that his Democratic opponent, Joseph R. Biden, Jr., would refuse to concede, leading to violence. "And when Donald Trump refuses to stand down at the inauguration, the shooting will begin," he said. "The drills that you've seen are nothing."

. . . Overall his tone was ominous: He warned, again without evidence, that "there are hit squads being trained all over this country" to mount armed opposition to a second term for Mr. Trump. "You understand that they're going to have to kill me, and unfortunately, I think that's where we're going," Mr. Caputo added.

In a statement on Monday, Mr. Caputo told The Times: "Since joining the administration, my family and I have been continually threatened" and harassed by people who have later been prosecuted. "This weighs heavily on us, and we deeply appreciate the friendship and support of President Trump as we address these matters and keep our children safe."[164]

Does that seem like an inversion of the world in which we're living? They say the right wing are the militants, and yet its Antifa and Black Lives Matter that have been so extremely violent, causing up to two billion dollars in property damage during the summer of 2020,[165] not to mention a number of deaths.

And for the claim that the protest at the Capitol on January 6, 2021 was an armed "insurrection," how is it that these Second Amendment-loving conservatives failed to bring a single one of the supposedly 400 million guns[166] that are currently in our society? Say what you will about conservatives, but they know you need to bring a gun to a revolution. The vast majority of those present intended January 6 to be a protest, a peaceful one, in my estimation, and not an insurrection. If someone was protesting and broke the law, that is a different issue and the law has to deal with them. However, my view has always been that for those that remained peaceful (this was the vast majority) and did not do anything illegal, then no enforcement actions should have been taken against them. I write this as bluntly as I just did because I am also a huge supporter of police and our military. I do think they are the finest among us.

When the question of whether there were irregularities in the November 3, 2020 election arises, I never hear the following line of questions asked of liberal activists. I imagine it would go something like this:

> **Interviewer:** Do you believe Donald Trump is an existential threat to our democracy, akin to Adolf Hitler?
>
> **Liberal Activist:** Yes, and in some ways he's worse than Hitler.
>
> **Interviewer:** Well, since that's what you believe, wouldn't you or any other activist be morally justified in doing whatever it took, including breaking election laws, to prevent this new Hitler from returning to power?
>
> **Liberal Activist:** (Stunned silence that goes on for several minutes.)
>
> **Interviewer:** (Breaking the silence.) If I believed I was keeping Hitler from maintaining power, I'd happily stuff a few ballot boxes. Wouldn't you?
>
> **Liberal Activist:** I think I hear my mom calling. She says I need to go and clean up my room.

That's at least the way I imagine it would go in my mind. The *New York Times* article on Caputo continued:

> While Mr. Caputo characterized C.D.C. scientists in withering terms, he said the agency's director, Dr. Robert R. Redfield, was "one of my closest friends in Washington," adding, "He is such a good man." Mr. Caputo is partly credited with helping choose Dr. Redfield's new interim chief of staff.
>
> Critics say Dr. Redfield has left the Atlanta-based agency open to so much political interference that career scientists are on the verge of resigning.

> The agency was previously seen as apolitical; its reports were internationally respected for their importance and expertise.
>
> Mr. Caputo charged that scientists "deep in the bowels of the C.D.C." walked "around like monks" and "holy men" but engaged in "rotten science."[167]

Now Caputo had a higher estimation of Dr. Redfield than I did, but both of us saw him as a good man, although I thought he did not have the strength to impose his will on the agency. I did grow to respect him and admire him as time went by, for he was a good person. I think it's probably difficult for the average person to understand the ways that government bureaucrats can short-circuit the plans of a leader who wants to reform the agency. However, I think the fact that Caputo had a close relationship with Dr. Redfield, and that Dr. Redfield himself was generating significant backlash at the CDC, should be taken as evidence that Caputo knew what he was talking about when he accused scientists at the CDC of walking around like "monks" and "holy men" who were engaged in "rotten science." The *Times* article next turned to Caputo's view of me.

> He fiercely defended his scientific adviser, Dr. Paul Alexander, who was heavily involved in the effort to reshape the C.D.C.'s Morbidity and Mortality Weekly Reports. Mr. Caputo described Dr. Alexander, an assistant professor at McMaster University in Canada, as a "genius."
>
> "To allow people to die so that you can replace the president" is a "grievous sin," Mr. Caputo said.
>
> A public relations specialist, Mr. Caputo has repeatedly claimed that his family and business suffered hugely because of the investigation by the special counsel, Robert S. Mueller III, into Russian interference in the 2016 presidential election. Mr. Caputo was a minor figure in that inquiry, but he was of interest partly because he had once lived in Russia, had worked for Russian politicians, and was contacted in 2016 by a Russian who claimed to have damaging information about Hillary Clinton.[168]

Nobody was a "minor figure" when it came to Mueller's Russian collusion investigation. Future historians will surely cite this shameful chapter in our nation's history as the time when our law enforcement and intelligence agencies sought to punish and paralyze those who had helped guide President Trump into the Oval Office.

Can I verify everything that Caputo said, and do I believe he was correct about all his claims? No, I cannot, and no, I do not. Yet Caputo was a sensible person and well thought out in all my dealings with him. I found him to be sensible in his thinking and views. I have nothing negative to say about Caputo. But the crazy thing about working inside the government and hearing how people talk when they think they're out of the limelight is that it becomes easy to believe anything is possible.

If you go back to the correspondence of our Founding Fathers, it becomes clear that they were excellent students of human nature. The reason they believed so deeply in transparency was because of their conviction that humans were guaranteed to do terrible things if they knew nobody was watching them. I imagine if some of the Founding Fathers were with me in that sixth-floor lunchroom in HHS and heard the plotting I heard, they would not have been surprised in the least. This is a claim by Mr. Caputo which I cannot verify:

> Some of Mr. Caputo's most disturbing comments were centered on what he described as a left-wing plot to harm the administration's supporters. He claimed baselessly that the killing of a Trump supporter in Portland, Ore., in August by an avowed supporter of the left-wing collective [Antifa] was merely a practice run for more violence.
>
> "Remember the Trump supporter who was shot and killed?" Mr. Caputo said. "That was a drill."[169]

And yet, as I write these words in September 2022, two years later, how do the claims of my good friend, Michael Caputo, strike me now?

I only need to rewatch the speech President Joe Biden gave on September 1, 2022, in front of Independence Hall in Philadelphia, bathed in blood-red lighting, to believe the leader of the free world was giving a dog-whistle to his left-wing allies to commit violence against their fellow citizens. Here is how the lighting was received by some Americans:

> Another theme in commentary about the speech, particularly from MAGA world, was the red background lighting during Biden's speech, which prompted a lot of colorful commentary and hyperbole, including numerous comparisons to Hitler and Satan. Fox News host Tucker Carlson referred to it as a "blood-red Nazi background." Nikki Haley said on Fox News that Biden, "looked like he was in the depths of hell," while Rudy Giuliani tweeted it looked like a "basement in hell."

. . . Many on the right also accused CNN of deliberately altering the color of the background during their broadcast of Biden's speech so that it became hot pink instead of fascist red:

[Tweet from Mia Cathell]—Watch the moment CNN slowly adjusts the camera settings when the blood-red lighting behind Dark Brandon is looking a little too authoritarian.

By the end, the stripes on the American flag are pinkish-purple . . .

A CNN representative told *Mediate* that the color shift was due to a technical issue with the pool video feed.[170]

It's easy to assume the worst of our political opponents. Let's assume the Biden people didn't intend to suggest violence, Nazism, or Satanism in their choice of lighting for the president's speech.

Maybe they simply wanted to make it fire engine red to suggest an emergency.

It's certainly possible, even likely.

Let's move on to the words of President Biden's speech. They're likely to give us greater clarity on the intentions of the Democratic Party. This is from early in the speech:

Too much of what's happening in our country today is not normal. Donald Trump and the MAGA Republicans represent an extremism that threatens the very foundations of our Republic.

No, I want to be very clear, very clear up front. Not every Republican, not even the majority of Republicans are MAGA Republicans. Not every Republican embraces their extreme ideology. I know, because I've been able to work with these mainstream Republicans.

But there's no question that the Republican Party today is dominated, driven, and intimidated by Donald Trump and the MAGA Republicans. And that is a threat to this country.[171]

Has any American president ever addressed part of the country as being "not normal?" That's the language of a bully. He ups the ante by calling conservative Republicans a "threat to this country."

How is that not a dog whistle to the violent, Antifa left?

What is the duty of any American when the president cites a "threat" to this country?

It's to stand up and take action, possibly even violent action. It's difficult to believe, but the calls for violence against a substantial part of the American electorate become even louder in Biden's speech and it was unnerving and very troubling.

> And here, in my view, is what is true: MAGA Republicans do not respect the Constitution. They do not believe in the rule of law. They do not recognize the will of the people. They refuse to accept the results of a free election, and they're working right now as I speak in state after state to give power to decide elections in America to partisans and cronies, empowering election deniers to undermine democracy itself.
>
> MAGA forces are determined to take this country backwards, backwards to an America where there is no right to choose, no right to privacy, no right to contraception, no right to marry who you love. They promote authoritarian leaders, and they fanned the flames of political violence that are a threat to our personal rights, to the pursuit of justice, to the rule of law, to the very soul of this country.[172]

How can President Biden stand up in front of the country and declare that conservatives do not care about the Constitution? After God and family, it's probably the thing in which they believe the most. And when is demanding transparency about the votes in an election a denial of elections? It seems to me to be a demand for the integrity of elections, not the denial of them.

And has any American president in a single paragraph dismissed their opponents as fanning the "flames of political violence" and claimed they are a "threat to our personal rights, to the pursuit of justice, to the rule of law, to the very soul of the country?"

In the final segment of his Facebook live post, taken while he was at his family's home in Buffalo and talking with some friends, Caputo said the following:

> Don't discount what I've said. I'm not going anywhere. They're not going to run me out. If the President asks me to leave, I will leave. I really want to leave. For some of you who know me, you know my health is failing. My mental health is definitely failing. I don't like being alone in Washington. The shadows on the ceiling in my apartment, those shadows are so long.
>
> Jerry, God I hope Durham is up to something. God, I don't know. But I know this. I'm going to keep my family safe and you need to keep your family safe, too. I'm going to shut off my Facebook soon because I don't want anybody leaking this to the media, and if they do, I don't care. But I'd just

rather not have that problem on top of this problem, with the attacks in the media today.

It's going to last for a couple days, and they'll find another shiny thing to go after. They want me gone. I'm not going anywhere. I don't care. They protested in front of my apartment in Washington yesterday. I'm not going anywhere.

The problem is in Washington I can't carry a gun. I can't be armed. So, into the breach, into the breach. I return to Washington after some hospital appointments tomorrow, and a funeral for a friend.

Marge, I don't care if you share. I'm going to turn this off for a little while anyway. But don't be upset when I turn off my account until next week. If you carry guns, buy ammunition, ladies and gentlemen, because it's going to be hard to get. It's a sad, sad situation.

I urge you, if you're in Buffalo, in New York, to look up the New York Watchmen. My friend, Charlie, has put it together. They're not a militia, they're just people who band together, they're friends of mine.

[There's some yelling off camera, Michael turns his camera around, we see a tree-lined Buffalo neighborhood street, and a car speeding off.]

Did you hear that? That guy drove by and called out "You should be ashamed of yourself!" Ladies and gentlemen, they're coming. Be ready. Be ready. And listen to the Grateful Dead. [Michael always did like to relax to his Grateful Dead music. Not what you expect from an uptight conservative, right?] Peace to all.[173]

This was the man in charge of communications at HHS for the COVID-19 crisis, the person at the center of the hurricane, with the ability to see all the moving parts, and he was saying the entire process was thoroughly corrupted.

The response of the mainstream media?

He's had a mental breakdown.

Well, maybe if the government of which you are a part is actively killing people through their advice of what you should and should not do (lockdowns, school closures, business closures, denial of early effective treatment, mandates with an ineffective harmful injection), it might affect your mental stability. How many days can you watch that destruction, and get up every day, cheerful to get to your job?

Of course, the outcome of this Washington, DC, tale was predictable, as on September 16, 2020, this appeared in the news:

> The Department of Health and Human Services has announced that Michael Caputo, its head of communications, will take a 60-day leave of absence "to focus on his health and the well-being of his family," days after posting a dark, conspiracy-laden video on Facebook in which he accused CDC scientists of plotting to undermine President Trump. Dr. Paul Alexander, a Caputo ally who tried to control Anthony Fauci's public statements and meddle with key CDC data, is departing the agency altogether, HHS announced.[174]

Do you see the narrative of this Shakespearean tragedy taking shape as the media wants you to understand it? Caputo was Othello, the noble Moor, with the tragic flaw of jealousy, which would lead him to murder his wife Desdemona, under the prodding of the evil Iago.

But instead of strangling his wife, Caputo was charged with the capital crime of criticizing CDC scientists.

However, his punishment would be a paltry two-month banishment from the government. He'd be back by November 16, 2020, better than ever!

It's just that the presidential election would've taken place nearly two weeks earlier.

Surely President Trump wouldn't need the services of one of his most loyal aides in the two months before the election, would he?

And what of the evil Iago in this story, me, who stood accused of the treasonous crimes of trying to "control Anthony Fauci's public statements and meddle with key CDC data"? All of this was a lie, and, if you are up to date on the machinations of CDC for the last eighteen months under Biden and Director Rochelle Walensky, did you not hear the CDC now admit they have made dramatic errors? Who has been vindicated?

I could not be allowed anywhere near the COVID-19 task force, even though I hadn't declared my real thoughts in a post on Facebook Live.

Was I a villain, a hero, or maybe something in between?

I'll let you be the judge.

I only ask that you let me present my evidence.

Maybe when I've finished, you'll understand why I've come to believe that no president, even Donald Trump, is genuinely allowed to lead this country.

But this is not a pessimistic tale.

If anything, it's a story of hope, because the enemy has shown their true face.

And they are not as strong as they would like you to believe.

Together, we are much stronger than the enemies of humanity.

* * *

If the epitaph on my grave reads, "He was a scientist who messed with the 'Holiest of the Holy,'" I will face my Creator standing tall with my head held high.

As much as I'm a man of faith, I'm also a man of science and do not let the two worlds mix. This was expressed to me as a young man by one of my professors who often said, "In God we trust, all others must bring data."

Do you know the inciting incident that forced me to resign?

Maybe you don't trust me to tell you the truth, so I'll give you the version promulgated by the *New York Times* in an article titled "Political Appointees Meddled in C.D.C.'s 'Holiest of the Holy' Health Reports," published on September 12, 2020.[175]

Now forgive me for a moment, but I thought this was science. Tell the truth and show your data. That's what I was taught. It's not like I was asking the Pope to hand over the Shroud of Turin for scientific experiments, which would destroy the holy relic. It's just common sense for somebody like me who's trained in evidence-based medicine to look at how the CDC's Morbidity and Mortality Weekly Reports (MMWR) were compiled, especially in the middle of an unprecedented global pandemic.

It's kind of like the Internal Revenue Service (IRS) auditing your tax return.

Yes, it's an inconvenience, but we do it.

As an evidence-based science expert, I'm kind of like that IRS agent who reviews your tax return. It's my job and my training to determine whether a practice is based on good science and optimal, trustworthy methodology.

If you don't have anything to hide, you shouldn't mind showing it, right?

Not with the CDC.

This is how the *New York Times* hysterically framed the issue, taking the CDC's side without the slightest pretense of journalistic objectivity:

> Current and former senior health officials with direct knowledge of phone calls, emails and other communications between the agencies said on Saturday that meddling from Washington was turning widely followed and otherwise apolitical guidance on infectious diseases, the Morbidity and Mortality Weekly Reports, into a political loyalty test, with career scientists framed as adversaries of the administration.

> They confirmed an article in Politico Friday night that the C.D.C.'s
> public morbidity reports, which one former top health official described on
> Saturday as the "Holiest of the Holy" in agency literature, have been targeted
> for months by senior officials in the health department's communication
> office. It is unclear whether any of the reports were substantially altered, but
> important federal health studies have been delayed because of the pressure.
>
> The reports are written largely for scientists and public health experts,
> updating them on trends in all infectious diseases, Covid-19 included. They
> are guarded so closely by agency staff members that political appointees only
> see them just before they are published. Health department officials have typ-
> ically only received notices of the titles of the reports.[176]

Maybe it's just me, but whenever I hear the complaint that somebody is
"meddling" in something, all I can hear is the voice of innumerable villains
in the old *Scooby-Doo and the Mystery Gang* cartoons, who when captured
would always say, "And I would have gotten away with it, if it wasn't for you
meddling kids!"

Back to the science.

I think it's fair to describe the situation in the following way.

CDC employees have traditionally generated their Morbidity and
Mortality Weekly Reports without any review from agency outsiders. They
have been especially shielded from the eyes of lawmakers and the American
taxpayers.

To me, it's like only allowing the Catholic Church to investigate claims
of priests molesting children.

Or that only police officers will be allowed to investigate claims of
police brutality.

Everybody understands the principle of oversight. Unbiased individuals
should oversee the important work of our public servants.

The CDC had no oversight of their Morbidity and Mortality Weekly
Reports, and, to this day, they still do not. This is how *Politico* reported the
dispute on September 11, 2020:

> In one clash, an aide to Caputo berated CDC scientists for attempting to use
> the reports to "hurt the President" in an Aug. 8 email sent to CDC Director
> Robert Redfield and other officials that was widely circulated inside the
> department and obtained by POLITICO.
>
> "CDC to me appears to be writing hit pieces on the administration,"
> appointee Paul Alexander wrote, calling on Redfield to modify two already

published reports that Alexander claims wrongly inflated the risks of coronavirus to children and undermined Trump's push to reopen schools. "CDC
tried to report as if once kids get together, there will be spread and this will
impact school reopening . . . Very misleading by CDC and shame on them.
Their aim is clear."[177]

Of course, I berated them. They were attempting to use children as a political weapon to harm President Donald Trump.

How are you supposed to act when children are being used as political
pawns?

Alexander also called on Redfield to halt all future MMWR reports until the
agency modified its years-old publication process, rather than a brief synopsis.
Alexander, an assistant professor of health research at McMaster University
near Toronto whom Caputo recruited this spring to be his scientific adviser,
added that CDC needed to allow him to make line edits—and demanded an
"immediate stop" to the reports in the meantime.

"The reports must be read by someone outside of CDC like myself, and
we cannot allow the reporting to go on as it has been, for it is outrageous.
It's lunacy," Alexander told Redfield and other officials. "Nothing to go out
unless I read and agree with the findings how they, CDC wrote it and I tweak
it to ensure it is fair and balanced and 'complete.'"[178]

Was I a heretic for saying somebody outside of the CDC needed to review
their work on the Morbidity and Mortality Weekly Report? This was the
report upon which all the decisions about opening up or shutting down
our society were being made. Moreover, my review of them showed how
flawed and corrupted they were, with half-baked incomplete reporting that
was slanted and pseudoscientific reporting that could not be published in a
proper scientific journal. You remember those stories in the media of people
dying in motorcycle or auto accidents, or of gunshot wounds, but they were
listed as COVID-19 deaths? Those were true. I'm not saying they were the
majority, but there was a lot that scientists at the CDC were getting blatantly wrong.

I'm happy to show my work (especially because *Politico* already published a screenshot of my email in their article), and I don't apologize for
any of it.

[S]o I request that CDC go back to that report and insert this else Michael [Caputo], pull it down and stop all reports immediately. CDC tried to report as if once kids get together, there will be spread and this will impact school reopening . . . that was the aim and that's how it reads and its disingenuous. Very misleading by CDC and shame on them. Their aim is clear. This hurts any President or administration. This is designed to hurt this President for their reasons which I am not interested in. I am interested in this or any President being served fairly and that tax payers' money not be used for political reasons. They, CDC, work for him. The public wants honesty and fair reporting so that they can be informed, not deceived.[179]

I am saying now publicly what I was saying then, privately. Scientists at the CDC and the NIH were actively pursuing a non-evidence-based set of policies with the aim of removing President Donald Trump from power. Moreover, they used their corrupted unscientific guidance documents and MMWRs to do this.

My boss, Michael Caputo, was doing his best to defend me, but we all know the die had been cast. Here's Caputo responding to questions about me.

Asked by Politico about why he and his team were demanding changes to CDC reports, Caputo praised Alexander as "an Oxford-educated epidemiologist" who specializes "in analyzing the work of other scientists," although he did not make him available for an interview.

"Dr. Alexander advises me on pandemic policy and he has been encouraged to share his opinions with other scientists. Like all his scientists, his advice is heard and taken or rejected by his peers," Caputo said in a statement.

Caputo also said that HHS was appropriately reviewing the CDC's reports. "Our intention is to make sure that evidence, science-based data drives policy through this pandemic—not ulterior deep state motives in the bowels of CDC," he said.[180]

But the CDC did not want evidence-based data to be used in their weekly reports because it would have supported Trump's claim that we needed to get the economy and schools moving again.

Can you understand the various crimes of which I was accused?

First, I wanted Dr. Anthony Fauci to stop his serial lying to the American people.

Second, I wanted the schools, and indeed, the entire American economy, to be opened up.

Third, I didn't want untested vaccines to be given to the American
public.

Fourth, I wanted outside, evidence-based experts to look at the work of
the CDC.

The bureaucratic deep state wanted me gone because I was a "meddler."
And they would make it happen.

* * *

After the news broke about me "meddling" with the CDC's Morbidity and
Mortality Weekly Reports, there was a media firestorm.

In addition to complaints in the media about my "meddling" with the
"Holiest of the Holy" of CDC reports, they published articles that refer-
enced my emails (but did not publish my emails) about school closures.

The White House called and told me I should hunker down, grant no
interviews, and not speak to anybody at my job, because it was unclear who
was leaking to the press.

I replied that, "I want to make sure everybody sees the entire communication, because I talk about natural immunity, and that's what we need
in the children."

But the White House disagreed with that approach as they felt it would
deflect from Trump's messaging. I was to simply go quiet. They told me,
"Just be quiet because in a few days something else will happen, and the
media's attention will move onto something else."

After about four days of this self-imposed silence (my best recollection)
in which nothing of note happened, I got a call from the human resources
division of HHS, and there were others on the call advising they were with
the legal department, and various other persons I did not know. They told
me I needed to resign by 4:00 p.m. that day, or they would fire me. "If you
resign by that time," it was explained to me, "we will put out a statement
that the White House no longer needs your services, and you will be given
two weeks' severance pay. If you don't, we will put out a statement that you
have been fired by Health and Human Services, and we will release unflat-
tering information about you."

As I recall, at one point I had people who informed me that they were
in the White House on one phone in one hand, and a woman from the

human resources division of HHS on another line on another phone in my other hand.

They told me the president was aware of the decision by the bureaucratic Deep State to terminate me because of my interactions with Dr. Fauci, but the president would not accept a resignation from me.

"I am sitting here with the decision-maker," I was told at one point by the people who identified from the White House, "and he is telling you not to resign. You are his appointee and they cannot fire you. Just hang on for a few more weeks (two weeks) and we will bring you into the White House as a special advisor to the president."

I put one phone up to the other so that the White House people could talk directly to the woman from HHS.

"Listen, Dr. Alexander," the woman from human resources said. "Health and Human Services has all the power here. You work for us. We can fire you if we want. The president has no power."

I continued to protest. "You mean, you're telling me that your president, the leader of our country and the free world and who leads the federal government, is telling you directly not to fire me and to stand down, but you're ignoring him? You're saying you don't care what the president says?"

"The decision has been made and you will be terminated, Dr. Alexander," the woman said. "We are offering you the chance to resign, and, as we said, the president has no power."

I thought about it.

Technically, the woman from HHS was correct. I worked for them, and they had the right to fire me. Remember, I had finally gotten a paycheck from the Department of Defense a few weeks prior because I was not being paid by the Deep State bureaucracy.

Even the president of the United States could not protect me.

I agreed to resign.

I drove to HHS, signed the necessary paperwork, and surrendered my laptop and government-issued cellphone and whatever else belonged to the government. Everything was checked off, and I left. All of this took place on the pavement outside of HHS at 200 Independence Avenue SW as I was told I would not be allowed to enter the building to hand back the property of the government.

I had been canceled by Dr. Fauci.

The HHS and COVID-19 task force did not want their work reviewed by an evidence-based researcher.

As Dr. Redfield had said about the six-foot social distancing rule, they were just making it up and seeing what they could get away with. I say this bluntly, while ensuring you understand, that, in my experience, Dr. Redfield and, I argue, Dr. Giroir, were the two scientists I worked with who were not trying to hurt Trump.

There was basically no science involved with the Deep State's decision-making, only what could hurt President Donald Trump's re-election chances.

* * *

You may not be familiar with the term "wrap-up smear," but I'll let Nancy Pelosi explain the concept from a June 22, 2017 weekly briefing on CSPAN. She is not saying that Democrats use this tactic, but claims it is something Republicans often use.

> You demonize—we call it the wrap-up smear. If you want to talk politics, we call it the wrap-up smear. You smear somebody with falsehoods and all the rest and then you merchandise it and then you write it and they'll say, see it's reported in the press that this, this and this, so they have that validation that the press reported the smear. And then it's called a wrap-up smear. Now I'm going to merchandise the press' report on the smear we made. It's a tactic.[181]

I'll let you come to your own conclusion as to whether that's a strategy commonly used by Republicans because they have so many good friends in the mainstream media, or whether this might be a tactic more suited and perhaps even better utilized by the Democrats.

However, at the very least, we have a good working definition of a "wrap-up smear," provided by the speaker of the US House of Representatives.

They waited until the middle of December, after Trump had "lost" the 2020 election, to use the "wrap-up smear" on me. This is how *Politico* characterized my actions in a December 16, 2020 article:

> "[I]t may be that it will be best if we open up and flood the zone and let the kids and young folk get infected" in order to get "natural immunity . . . natural exposure," Alexander wrote on July 24 to Food and Drug Administration Commissioner Stephen Hahn, Caputo, and eight other senior officials . . .
>
> Alexander was a top deputy of Caputo, who was personally installed by President Donald Trump in April to lead the health department's

communications efforts. Officials told Politico that they believed that when Alexander made recommendation, he had the backing of the White House.

"It was understood that he spoke for Michael Caputo, who spoke for the White House," said Kyle McGowan, a Trump appointee who was CDC chief of staff before leaving this summer. "That's how they wanted it to be perceived."[182]

Let's get right to it, shall we? If this is supposed to be about the emails I wrote, specifically the ones in which they lifted four simple words, "We want them infected," as the cross on which my career would be crucified, I think it's important that you read the emails in their entirety. Here is one of the emails I sent on July 4, 2020, at 11:50 a.m. to Michael Caputo and others.

The ironic thing about these emails is I was agreeing with what Dr. Anthony Fauci had just said on a *Fox News* interview. The subject line of my email read: "Fauci Says Now Today on News that Vaccine Will Not Get Us to Herd [Immunity] alone . . . That Intuitively Means We Need Infected People."

First, these scientists and experts are just confusing and frightening the people. And they can't get their statements straight.

If Fauci now says vaccine won't help fully, then this is why we need a portion, a large portion of the population to be infected, recover, and have antibodies.

To get antibodies one needs to be infected first, as we know. It can't be done with lockdowns, etc. And each time you lock down and then open up, cases will spike.

I am trying to be as clear as possible:

We have never ever done this by locking down a healthy population . . . we protect the at-risk and let the rest of the well society go on . . . and face the pathogen. We will know in years the impact of this . . . but there is no evidence that locking down a healthy, well society, a well group of people actually works. Never. The issue is we want to help protect the elderly and those who are at risk. This is a basic need. So we vaccinate the 'well' and healthy and younger and get them to be infected and develop antibodies and it is them whose immunity will protect the vulnerable among us. We do this all the time. When we have a vaccine, we get partial immunity that way combined with immunity from those who get infected and recover due to exposure, and then we get to that herd number threshold where the pathogen is boxed in and can't spread and goes away. Sometimes we do herd with full vaccination e.g. we did this mostly with smallpox in the 1970s/1980s.

Point being, if Fauci is saying today the vaccine won't get us to herd [immunity] (I have no idea why he runs to the camera to tell the world each thought he has and often they are wrong sided and contradictory and I am sure the WH did not clear that as it now concerns people) then to get there we need persons in the population to augment this to get to that herd number of 60% (some now say only 40% of the population is needed to be immune for herd [immunity] to be established and is being debated) . . . this means exposure to the virus, exposing younger, well people to the virus and they develop immunity.

Thus the discussion has to be urgently, how do we protect the vulnerable, while letting the rest of the society free . . . the infants and the children who are at low or no risk, the teens who are also at no risk, the young people at very small risk of severe illness, younger adults at low risk and folks with no conditions, etc . . . allowing them to spread it and be infected . . . kind of like measles parties. I am trying to tell you that Fauci is undercutting (thwarting all efforts to deal with the virus in a positive way) the message or not working with us to package it so we can get the population informed and on board, he is frightening people . . . and at the same time, we need to consider this closing back down or slowing down opening up . . . I do not agree . . . if there are upticks in cases due to testing and more mingling etc., loosening restrictions, the issue is are these going onto severe illnesses or you just feel ill for a few days and need chicken soup. If you are not gravely ill (or even have no symptoms and don't feel anything), then we may need to consider letting our society loose, if what Fauci says is so.

There is no other way, we need to establish herd [immunity], and it only comes about allowing the non-high-risk groups [to] expose themselves to the virus. PERIOD. We continue the public health messages of proper hygiene, handwashing, protecting elderly at your homes and nursing care facilities, social distancing and so on . . . but we go on and let our societies open up fully NOW. If the hospitalizations occur in [the] young and they get severe illness, then that's a different story and we will then have a huge mess on our hands and real nightmare. But that is not the case now.

Dr. Paul E. Alexander, PhD
Senior Advisor to the Assistant Secretary
For COVID-19 Pandemic policy
Office of the Assistant Secretary of Public Affairs (ASPA)
US Department of Health and Human Services (HHA)
Washington D.C.[183]

Did I sound like a crazy person? Or more than two years after this series of events, do I sound like just about the only rational person on the COVID-19 task force? At 1:44 p.m., I sent a follow-up email to the same group.

Fauci, who is for lockdowns and the like, is now actually clarifying things by his statement on Fox that we won't get to herd [immunity] with a vaccine . . . he is actually helping the administration.

He is actually one of the architects of the lockdown, but is actually saying now indirectly, that [the] vaccine can't do it, so we will need the population to get infected and develop antibodies. The issue is six months in, only about 5% of population show antibodies and this means it will take 7-9 years to get 60% herd [immunity] by way of lockdown, partial lockdown, open up, close back down, etc.

My view, we open up fully as described below, protect the vulnerable, make sensible decisions, and allow the nation to develop antibodies. Infants, kids, teens, young people, middle-aged with no conditions, etc, have zero to little risk . . . so we use them to develop herd [immunity] . . . we want them infected . . . and recovered . . . with antibodies . . . hospitals are NOW geared [up], PPE in place, ICU beds are on the ready, doctors and nurses alert, the syndrome is crystalized . . . etc. Only if the young who are getting infected with the increased testing and relaxed controls now . . . if they show serious illness needing ICU and oxygen, and die, then we know this virus has mutated lethally and attacking the usual healthy in a society and this is dangerous and I don't think so . . . god forbid this ever happens . . . data does not show this. Data shows now that only 3.5% of deaths now are in persons young than 44 years . . .

You can't discount the devastation of lockdowns as people lose independence, homes, jobs, hope, kill themselves, drink and use illicit drugs and die to substance abuse deaths . . . this is a fact . . . more will die due to the indirect effects of COVID . . . these deaths of despair . . . if the young, healthy, well among us can face the pathogen and develop herd [immunity], why can't we do this? Prior has not worked . . . you can't lock down and reopen and if spike, close down again . . . the virus will never go away . . . it will lurk . . . we got to face it once and for all . . . while pushing hard for vaccine and therapeutics . . . all at once . . .

And stop Fauci from talking . . . he is confusing people . . . he flip flops the message too much . . . and the result now is he is not credible . . . I talk to lots of people and read the IT world. Others are . . . he is not. His messaging is not consistent, so you don't know what the best science is advocating.

Dr. Paul E. Alexander, PhD
Senior Advisor to the Assistant Secretary
For COVID-19 Pandemic policy
Office of the Assistant Secretary of Public Affairs (ASPA)
US Department of Health and Human Services (HHA)
Washington D.C.[184]

There you have the complete story of the "wrap-up smear" about me. They tried to make me into the monster, the villain to the saintly Dr. Fauci.

When my time comes, I will stand proudly before my Maker, and defend everything I tried to do during the COVID-19 crisis.

If I am granted access through the pearly gates, I wonder if I will find Dr. Fauci waiting for me among the mansions of the Lord, or will the angels inform me Dr. Fauci has been sent to "that other place?"

* * *

Did President Trump lose the 2020 election?

I can't give you a definitive answer.

It seems to me there are four areas of concern.

The first is what happened on election night. After the election, Dr. Peter Navarro, the White House director of Trade and Manufacturing Policy, one of the genuinely good guys of the Trump administration (who is now being persecuted by the January 6, 2021 Commission), put together a series of reports about election irregularities.

At the stroke of midnight on Election Day, President Donald J. Trump appeared well on his way to winning a second term. He was already a lock to win both Florida and Ohio; and no Republican has ever won a presidential election without winning Ohio while only two Democrats have won the Presidency without winning Florida.

At the same time, the Trump-Pence ticket had substantial and seemingly insurmountable leads in Georgia, Pennsylvania, Michigan, and Wisconsin. If those leads held, these four battleground states would propel President Trump to a decisive 294 to 244 victory in the Electoral College.

Shortly after midnight, however, as a flood of mail-in and absentee ballots began entering the count, the Trump red tide of victory began turning Joe Biden blue. As these mail-in and absentee ballots were tabulated, the

President's large leads in Georgia, Pennsylvania, Michigan, and Wisconsin simply vanished into thin Biden leads.[185]

This is simply a factual statement of what happened on election night 2020. Perhaps there's a reasonable explanation. However, it must certainly be one of the strangest reversals ever on Election Day. Dr. Navarro continued to explain the Election Day weirdness:

At midnight on November 3, and as illustrated in Table 1, President Trump was ahead by more than 110,000 votes in Wisconsin and more than 290,000 votes in Michigan. In Georgia, his lead was a whopping 356,945; and he led in Pennsylvania by more than a half million votes. By December 7, however, these wide Trump leads would turn into razor thin Biden leads—11,799 votes in Georgia, 20,682 votes in Wisconsin, 81,660 votes in Pennsylvania, and 154,188 votes in Michigan.[186]

There were many calls to investigate these irregularities. To my knowledge, no court has actually investigated these claims of widespread voter fraud, normally claiming they lacked jurisdiction over the matter. This abandonment by the judicial system caused Trump and his followers to call for a protest on January 6, 2021, demanding that the votes of certain states be sent back to those legislatures to conduct a ten-day review to confirm the results.

In the ensuing violence of the January 6 Capitol protest, claims were made that those who engaged in this violence were in actuality federal agents engaged in violent provocation designed to discredit the movement.

In January 2022, Senator Ted Cruz asked a series of questions to the FBI about the events of January 6, 2021, which went unanswered.

"How many FBI agents or confidential informants actively participated in the events of January 6?" Texas Republican Sen. Ted Cruz asked the Executive Assistant Director for the FBI's National Security Branch Jill Sanborn.

"I can't go into the specifics of sources and methods," Sanborn said.

"Did any FBI agents or confidential informants actively participated in the events of Jan. 6, yes or no?" Cruz pressed.

"I can't answer that," Sanborn said, despite the New York Times revealing in September the presence of federal agents within the crowd that stormed the Capitol. Earlier this month, Newsweek revealed in a blockbuster story the presence of secret commandos with "shoot to kill" authority.

Did any FBI agents or confidential informants commit crimes of violence on Jan. 6?" Cruz followed up

"I can't answer that," Sanborn repeated.

"Did any FBI agents or FBI informers actively encourage and incite crimes of violence on Jan. 6" Cruz asked.

"I can't answer that," Sanborn said for a third time.

Cruz went on to ask the agency executive about Ray Epps, a suspect captured on tape encouraging people to enter the Capitol who disappeared without explanation from the FBI's Capitol Violence Most Wanted List last summer, according to an October report from Revolver.

Sanborn said she was "aware of the individual" but lacked "specific background."[187]

We commonly say our politicians and bureaucrats lie, but when testifying under oath to Congress they generally use an evasion and misdirection strategy. However, when the questions are as skillfully asked, as they were by Senator Cruz, they have no wiggle room, and generally decline to answer.

Their failure to answer should tell us everything we need to know. And again, I need it crystal clear that I am one of those who support law enforcement, our police, and our military. I think that the vast number of rank and file FBI agents are credible people.

* * *

The second curious thing during the 2020 elections that should concern us is the actions of Facebook billionaire Mark Zuckerberg.

This is an article from National Public Radio (NPR) on December 8, 2020 with the title, "How Private Money from Facebook's CEO Saved the 2020 Election." (I thought you were an insurrectionist if you believed there was the slightest thing wrong with our elections. If nothing was wrong, or ever had been wrong, why would it need to be saved by Mark Zuckerberg's hundreds of millions of dollars?) Here is a section of the article:

With a tight budget and little help from the federal government, Chester County applied for an election grant from the Center for Tech and Civic Life, a previously small Chicago-based nonprofit that quickly amassed hundreds of millions of dollars in donations to help local election offices—most notably $350 million from Facebook CEO Mark Zuckerberg and his wife, Priscilla Chan.[188]

It's remarkable to me how shamelessly these media people lionize Zuckerberg's infusion of $350 million into the 2020 election. Do they want us to genuinely believe hundreds of millions of dollars were spent simply to help "democracy" rather than their favored candidate, Joe Biden?

Then there was the remarkable appearance by Zuckerberg on *The Joe Rogan Experience* podcast on August 25, 2022, in which Zuckerberg told Rogan that in the weeks leading up to the 2020 election, the FBI visited Facebook and told them to be aware of a dump of "Russian misinformation." Lo and behold, the Hunter Biden laptop story magically dropped a few days later and Facebook decided it must be what the FBI was talking about when they warned about "Russian misinformation." From a BBC article about the interview:

> Zuckerberg told Rogan: "The background here is that the FBI came to us—some folks on our team—and was like "hey, just so you should know, you should be on high alert. We thought there was a lot of Russian propaganda in the 2016 election, we have it on notice that basically there's about to be some kind of dump that's similar to that."
>
> He said the FBI did not warn Facebook about the Biden story in particular—only that Facebook thought it "fit the pattern."[189]

Again, are they asking us to believe all of this was some unfortunate misunderstanding between the FBI and Facebook? To me it seems like the situation was perfectly designed in a way to deflect blame from either side, or, as they would say in the intelligence community, for "plausible deniability."

The third thing you might not know about Mark Zuckerberg and the 2020 election is that he gave his personal phone number to Dr. Anthony Fauci. I was the senior pandemic advisor and Zuckerberg never gave me his personal phone number. Zuckerberg never gave Caputo his phone number. Caputo never spoke to Zuckerberg as far as I know. Why? Why Dr. Fauci then? In terms of the pandemic response, no one in the US government was more important to the response as was Caputo. He was central to it. I know who we dealt with daily, and I know the issues that passed through the offices and across the desks.

This is from an article in the *New York Post* on September 9, 2022, describing this unusual collaboration between Mark Zuckerberg and Dr. Anthony Fauci:

The ugly picture of collusion between the feds and social media platforms around COVID just got a whole lot uglier.

Recent filings from a lawsuit by the Louisiana and Missouri attorneys general against the Biden administration reveal Facebook head Mark Zuckerberg gave Dr. Anthony Fauci his personal phone number shortly before the platform started to crack down on alleged COVID misinformation.

Is *this* how *The Post*—and many others—got banned, throttled or labeled as purveyors of misinformation for merely raising the possibility (as we did in a prescient February 2020 op-ed) that COVID originated from an accidental leak in Wuhan?[190]

It's amazing how, in the election season of 2020, Mark Zuckerberg seemed to find himself the subject of so many interesting stories.

With the cover-up of the Hunter Biden laptop story, the hundreds of millions of dollars spent during the 2020 election, and the direct line to Dr. Anthony Fauci's personal phone, the question is whether Zuckerberg "saved" democracy (as his supporters claim), or whether he bought his preferred president the job.

* * *

The third thing that should concern us about the election of 2020 actually had its roots in a 2016 statement by the majority leader of United States Senate, Chuck Schumer of New York, in an interview with Rachel Maddow of MSNBC, broadcast on January 3, 2017:

> **RACHEL MADDOW:** Let me ask you. I don't know if you've seen this. I don't want to blindside you with this. This is the latest statement, latest tweet as you were just saying—President-Elect's latest unsolicited pronouncement on the intelligence community. This was his tweet just a little while ago last night. You can see the scare quotes here. "The 'intelligence' briefings on so-called Russian hacking was delayed until Friday. Perhaps more time needed to build a case. Very strange."
>
> We're actually told—intelligence sources tell NBC News that since this tweet has been posted, that actually, this intelligence briefing for the President-Elect was always planned for Friday. But he's taking these shots, this antagonism. He's taunting the intelligence community.

> **SENATOR SCHUMER:** Let me tell you, you take on the intelligence community, they have six ways from Sunday at getting back at you. So, even for a practical, supposedly hard-nosed businessman, he's being really dumb to do this.[191]

Even though I was an important government official, I may have forgotten the lines of authority in the United States Constitution. It is my understanding that the intelligence services work for the president, not the other way around.

But perhaps it's more easily understood if we take the perspective that everything we've suffered since January 2020 was the result of the failures of the intelligence agencies, which they have been desperate to cover up.

Trump, and the millions of people who raised similar questions, were just in the way of their plans.

It was nothing personal.

It was business.

* * *

The fourth important thing to consider is that all of the economic pain, suffering, and death might have been the direct result of decisions made by the intelligence services of our country.

On January 10, 2022, the investigative group Project Veritas, led by the charismatic James O'Keefe, released an astonishing set of documents. The documents detailed our government's response to proposed gain of function research on bat coronaviruses.[192]

To set the stage, on January 19, 2018, the Defense Advanced Research Projects Agency (DARPA) released a "Broad Agency Announcement" about "PREventing Emerging Pathogenic Threats (PREEMPT)."[193] EcoHealth Alliance, under the leadership of Dr. Peter Daszak, submitted a proposal. They wrote:

- Security concerns across Asia make the region a potential deployment site for US warfighters. Troops face increased risk from SARSr-CoVs, which are shed via urine and feces as bats forage at night.
- Our work in Yunah, China shows that: 1) bat SARSr-CoVs exist that can infect human cells, and are not infected by monoclonal or vaccine treatments; and 2) bat SARSr-CoV host-jump into local

human populations is frequent. These viruses are therefore a clear and present danger to US defense forces in the region and global health security.

- Our goal is to analyze, predict, then "DEFUSE" the spillover potential of novel bat-origin high risk SARSr-CoVs in Southeast Asia and across the virus's distribution. This will safeguard the US warfighter, reduce risk for local communities and their livestock, improving food and global health security.

- Our strategy is based on immune parameters that are found across all bat genomic groups. If successful, the DEFUSE approach can be adapted to other MERV-CoV in the Middle East, other SARSr-CoVs in Africa, and other bat origin viruses (e.g., Hendra, Nipah, Ebola, Marburg viruses.)[194]

The plan was relatively simple: identify potentially harmful bat coronaviruses, model their likely evolution by genetically manipulating them in humanized mice, and then create vaccines that could be sprayed on the bats, or which could be sprayed as a mist in their caves that the bats would inhale.

Section II of the report detailed how this was to be done:

> Our goal is to defuse the potential for spillover of novel bat-origin high zoonotic risk SARS-related coronaviruses in Asia. In TA1 we will intensively sample bats at our field sites where we have identified high spillover risk SARSr-CoVs. We will sequence their spike proteins, reverse engineer them to conduct binding assays, and insert them into bat SARSr-CoV (WIV1, SCHO14) backbones (then use bat-SASr-CoV backbones, not SARS-CoV, and are exempt from dual-use and gain of function concerns) to infect humanized mice and assess capacity to cause SARS-like disease. Our modeling team will use these data to build machine-learning genotype-phenotype models of viral evolution and spillover risk.[195]

DARPA, which is often portrayed as the craziest group of scientists to ever work in government, was so terrified of this research that they rejected it, citing many potential safety risks.

However, this research was eventually funded by Dr. Anthony Fauci and his agency, NIAID. An article from January 2022 from *Fox News* laid out the allegations:

According to the timeline of events laid out by Baier, Fauci was told on January 27, 2020 [about seven weeks before the nationwide lockdown] that his NIAID [National Institute of Allergy and Infectious Diseases] had been indirectly funding the Wuhan lab through EcoHealth Alliance—a US based scientific nonprofit that had been working with novel coronaviruses.

On January 31, Dr. Kristian Andersen, a noted virologist at Scripps lab, privately told Fauci that after discussion with colleagues some of COVID-19's features look possibly engineered and that the "genome is inconsistent with expectations from evolutionary theory."

Andersen added that the situation needed to be looked at more closely, at which point Fauci organized an all-hands-on deck conference call with colleagues where he was told that risky experiments with the novel coronavirus may not have gone through proper biosafety review and oversight.[196]

As early as January 27, 2020, Dr. Anthony Fauci may have known his agency was responsible for the impending terror and death of COVID-19, and yet acted as if the pathogen had caught him by surprise.

And what are we to make of Dr. Peter Daszak and his organization, EcoHealth Alliance?

According to Dr. Andrew Huff, the former associate vice-president of EcoHealth Alliance, Daszak was actively courting the Central Intelligence Agency (CIA) to keep the organization afloat. A January 20, 2022 article from the *National File* reported on the claims Huff made about his former employer and the CIA in a series of tweets on January 12, 2022:

For the record: In 2015 Dr. Peter Daszak stopped me as we were leaving work late at night and asked me if he should work with the CIA. I was shocked given my experience in security. Over the next 2 months he gave me updates on 3 separate occasions about his work with the CIA.

When he asked me the question, I stated, "Peter, it never hurts to talk with them and there could potentially be money in it." Meanwhile, I was cringing that he told me this, in a non-classified setting (a SCIF [sensitive compartmented information facility]), to a person that was not "read-in," and to an uncleared person (me).

Then, over the next two months at the break area while getting coffee, or between meetings, he stated that they were interested in the places that we were working, the people involved, the data we were collecting, and that the work with them was proceeding.[197]

Could it be that Daszak's plan, the one which was so dangerous that the scientists at DARPA turned it down, had been picked up by the CIA? The allegations of Andrew Huff continued.

> Prior to the public statement earlier that Morning, Huff took to Twitter and claimed "members of the US government IC [Intelligence Community] have been harassing me, broke into my house, stole hard drives, and installed electronic surveillance devices throughout my house . . .
>
> In a separate series of posts, Huff said that he "wouldn't be surprised if the CIA/IC community organized the COVID coverup acting as an intermediary between Fauci, Collins, Daszak, [Dr. Ralph] Baric [of the University of North Carolina], and many others. At best, it was the biggest criminal conspiracy in US history by bureaucrats or political appointees."[198]

Do all trails lead back to the intelligence community, perhaps working hand in latex glove with the pharmaceutical companies?

It would be consistent with what I saw of Operation Warp Speed in terms of heavy military involvement on the seventh floor of HHS.

I realize there are several possible scenarios.

It could be that there was hostility to Trump *and* there was this accidental release of COVID-19 from the lab in Wuhan, China, and that those who were hostile to Trump saw this as an opportunity to damage him.

It could be that the release of COVID-19 was intentional, designed to damage Trump as a prelude to elements of the government seeking to gather additional powers for themselves. This is a common tactic of authoritarian governments seeking to create a crisis, and thus benefitting from the ensuing fear. Every government seeks ways to amass power and, once they have it, will maximally abuse it (e.g., the COVID emergency declaration powers).

It could be that the scientific community was genuinely in the dark about COVID-19, and although they may have engaged in measures that may not have worked as intended, were genuinely trying to do the best for the benefit of the public.

However, as the government's own evidence-based medicine expert on COVID-19, I can't tell you which scenario is the most likely.

I have my suspicions but lack the information necessary to come to a definitive conclusion.

Where does that leave us?

Is this an unproven hypothesis, like the man accused of murder, for which the state lacks the evidence to prove guilt to a jury's satisfaction, and thus we must proclaim him innocent?

I think not.

When one individual faces the awesome power of the state, I think it is correct to force the state to prove the allegations beyond a reasonable doubt.

But the situation is different when we speak of the government, or elements of it.

The government has all the power.

Therefore, when credible allegations arise about government misconduct, we must demand the government prove their innocence beyond a reasonable doubt. It is a fair standard.

In the absence of such a defense, we must assume the government is guilty of some crime.

What did I see during my time as senior pandemic advisor?

I saw strategies put in place that did not follow the science, specifically for lockdowns, school closures, denial of early treatment, and vaccine development.

I saw a Deep State bureaucracy that was hostile to President Trump and actively sought to make his presidency a failure. I saw alphabet health agencies and regulators working against the best interests of America and Trump.

I saw a president who was surrounded by people who were not genuinely loyal to him, and wanted to benefit themselves and serve the bureaucracy, rather than serve the American people.

I saw scientists who raised reasonable questions about COVID-19 origins, or the lockdown strategies, unfairly attacked and vilified in a manner foreign to the principles of scientific debate.

Like you, I watched such bizarre extreme reported news like the parade of troubling news about the efforts of billionaires to intervene in the election process, even threats from leading politicians of incurring the wrath of the intelligence community, and the remarkable reversal of vote tallies from the night of November 3, 2020 into the early morning hours of November 4.

Do I think President Trump lost the 2020 election?

No.

I think we witnessed a Presidential Takedown, an alliance between democrats, the bureaucratic Deep State, RINOs, legacy media, Hollywood, academia, medical journal publishers, alphabet agencies such as the CDC, NIH, FDA, and NIAID (and their top officials), and the pharmaceutical industry, as well as others, to not only drive President Trump from power, but to forever change the relationship between the government and its citizens.

In this book I have made my allegations.

The burden now shifts to the government to prove beyond a reasonable doubt that they acted in good faith during the COVID-19 crisis. Trust in the government needs to be restored.

And in the meantime, how are we to view our government?

In a high-profile murder case, the judge may determine that the suspect poses such a risk of harm to the public that the accused must remain in jail until the facts can be established.

More than six million people died during the COVID-19 crisis, in addition to the harm done to school children, countless business owners, and those who died from drug overdoses due to depression, undiagnosed illnesses because of the lockdown, and countless suicides. Many American children committed suicide due to the lockdowns and school closures, not the virus.

Is it time for us to lock up the members of our public health establishment until they answer our questions? Is it time based on proper public and legal inquiry, if it is shown that they were at fault and caused deaths by their policies, to impose harsh punishment? If so, I say lock them all up! All of them who caused deaths needlessly.

I have been clear in stating my opinion that the events of the COVID-19 pandemic resulted in President Trump not winning reelection in 2020. I place the blame for this unprecedented interference in our political system at the feet of Anthony Fauci, Deborah Birx, the NIH, CDC, and the WHO.

What I witnessed was an insurrection, plain and simple, against the American republic.

I have told you what I saw and how I interpreted such events. One may say I have given a first draft of history, with all of the attendant drawbacks and strengths. Perhaps some of my criticisms have been too harsh, or not harsh enough. I've also been clear I am still subject to security oaths, so if there is opinion I render, and you ask, "Paul, why do you think that?" I may not be able to give you a satisfactory answer.

However, whatever I or other members of the COVID-19 task force may say, it is others, scientists, historians, maybe the courts, and eventually public opinion, which will define the truth of what happened.

CHAPTER SEVEN

The Path Forward Out of the COVID Mandates, Restrictions, and Lockdown Lunacy

It is now clear[199] that the COVID-inspired lockdowns,[200] school closures,[201,202] mask mandates,[203] and all of the COVID-19 pandemic restriction policies that societies were subjected to over the last two and a half years have all failed catastrophically.[204]

Our societies were decimated with lockdown policies that were crushing, and the shocking reality is that none worked. Even the efforts to mislead societies about the superiority[205] of natural immunity over vaccinal immunity failed and the science was disregarded.

Governments enacted devastating and restrictive COVID policies on their societies with irrational, illogical, unscientific, specious, and unsound arguments, which were often lacking any evidentiary basis.

It has been estimated that it will take decades to recover from what our governments did. The costs have been staggering in terms of damage to mental health of children,[206] the consequential rise in hunger and poverty, the crushing effects on economies,[207] the loss of education, escalated costs to health care, the delayed and canceled care for non-COVID illnesses, and the impact on crime.

Tens, if not hundreds, of thousands (and potentially millions) were denied treatment for other medical conditions, and the excess-mortality data is indicating the catastrophic effects of this. The same number again was also denied early outpatient treatment that was known to be safe, effective, not cost prohibitive, and already approved by the regulators.

Lockdowns Did Not Protect

Lockdowns did not protect the vulnerable but harmed them, shifting the morbidity and mortality burden to the underprivileged who could not afford to "shield." We instead locked down the healthy, while at the same time failing to properly protect the actual group that lockdowns were proposed to protect—the high-risk vulnerable and elderly. We shifted the burden to the poor (women, minorities, and children) and caused catastrophic consequences for them, as they were in the least-viable position to lock down. Those in the "laptop, Zoom, café latte" class seemed unperturbed by the larger societal ills and consequences and called for more, longer, "harder" lockdowns.

In a sense, what we have done is perverse and sickening, with calls (then and even now) from those more affluent groups to maintain lockdowns as they have "settled" into a comfortable cadence and flow, a more "structured" life now fulfilled with Uber Eats and Amazon. They can walk their dogs, tend to the garden, and go for coffee as they wish. They could slip in a mini-vacation when they wanted.

Wealth disparities placed those who were more vulnerable economically in a difficult position in terms of sheltering from the pandemic. It left them highly exposed, and COVID decimated them. COVID-19 emerged early on to be a disease of great disparity, and we learned a few weeks out that it was amenable to risk stratification and that baseline risk was prognostic on severity of outcomes and mortality. A steep age-risk curve emerged, and this underpinned our calls for a "focused" protection as advocated in the Great Barrington Declaration.[208]

Lockdowns badly harmed the elderly, leaving them confined in their nursing homes and extending the window of exposure to the virus for them. And they were subject to repeated exposure from staff that brought the pathogen into the confined settings, which drove hospitalizations and deaths.

Lockdowns thus reduced the movement of younger, low-risk persons (who were most able to handle the pathogen immunologically) to the same level of movement and mobility as the elderly, higher-risk persons, and thus equalized the chance of infection between the low-risk and high-risk, the young and the old. But while lockdowns equalized the risk, the way out was to maintain a risk differential as we protected the vulnerable. Failing to do so was catastrophic, as it denied movement toward population immunity in most instances.

The lockdowns were really the key feature of global governments' COVID pandemic actions and really worked to disable and crush societies. They turned out in all locations and nations to have been counterproductive and unsustainable. They were meritless and unscientific. There was no good reason and no sound justification to harden lockdowns and to keep them going after we quickly learned in the spring of 2020 how to manage COVID and who was the at-risk group.

Data That Should Not Be Ignored

These unparalleled policy actions were enacted for a virus whereby the median/mean age of death began in February 2020 at about eighty-two to eighty-three years of age and has remained so until September 2022. This was similar to, or greater than, the typical life expectancy of approximately seventy-nine to eighty years in most nations. If you were high risk and did succumb to COVID-19, you were at an almost 100 percent chance of having lived past your national life expectancy. COVID-19, despite what the media wants you to believe and has stated for thirty months now, has not shortened lives in general.

Do we have sound evidence of risk? Well, the reality is that there was so much societal damage for a virus with an infection fatality rate (IFR) roughly similar (or likely lower once all infection data are collected) to seasonal influenza. Stanford University's Dr. John P. A. Ioannidis identified thirty-six studies (forty-three estimates) along with an additional seven preliminary national estimates (fifty pieces of data) and concluded that among people under seventy years old across the world, infection fatality rates ranged from 0.00 percent to 0.57 percent, with a median of 0.05 percent across the different global locations (with a corrected median of 0.04 percent).[209]

Moreover, the IFR has been shown to be near zero for children[210] and young adults. Evidence showed us early on that while anyone is at risk of being infected, "there is more than a thousand-fold difference[211] in the risk of death between the old and the young."

Omicron Presents as a Common Cold to Most

Omicron sub-variants BA.4 and BA.5 present as a common cold to most. While recognizing that it could be challenging to the high-risk elderly (as are common colds and flu), commonsense precautions need to be taken. We argue that the pandemic is over, and therefore the emergency declarations must be ended.[212] These include all mandates and all restrictions.

Omicron is mild enough that most people, even many high-risk persons, can sufficiently cope with the infection. As mentioned, omicron infection is no more severe than seasonal flu, and generally less so. We have learned much about the utility of inexpensive supplements like vitamin D to reduce disease risk, and there is a host of good therapeutics available to prevent hospitalization and death, should a vulnerable patient become infected. And for younger people, the risk of severe disease—already low before omicron—is <u>minuscule</u>.

No Reason Now for Keeping State of Emergency

Given that omicron, with its mild infection, is running its course to the end, there is no justification whatsoever for maintaining the emergency declaration status. The lockdowns, personnel firings and shortages, and school disruptions have done at least as much damage to the population's health and welfare as the virus.

The state of emergency is not justified now, and it cannot be justified by any fears of a hypothetical recurrence of some more severe hypothetical infection at some unknown point in the future. If such a severe new variant were to occur—and it seems unlikely from omicron—then that would be the time to discuss a new declaration of emergency[213]. However, caution is urged. This pandemic could have ceased and settled into an endemic phase with mild seasonal circulation.

However, with the use of non-sterilizing, non-neutralizing, antigen-specific vaccines that induce vaccinal antibodies that do not eliminate the virus (that have become largely resistant to the potentially neutralizing vaccinal antibodies), we could see the maintenance of the pandemic with the selection of infectious variant after infectious variant. This is an ongoing concern given the emergence of viral immune escape and antibody-dependent enhancement of infection (ADEI) and disease (ADED), as well as original antigenic sin (immune priming, immune fixation, and prejudicing to the initial prime); we could be maintaining this pandemic.

The Vaccine Is Driving Infection in the Vaccinated

The evidence is now clear that it is the COVID gene injection vaccine platform that is driving massive infectious pressure in the midst of sub-optimal vaccinal antibodies that do not cut the chain of transmission. It is the vaccine that is driving infection in the vaccinated. This continues to be a recipe for disaster and either the infectious pressure must be cut or the injection must be stopped (and, if used, targeted only for the high-risk vulnerable).

For additional reading, please refer to Liu et. al. "An Infectivity-Enhancing Site on the SARS-CoV-2 Spike Protein Targeted by Antibodies"[214] and Fantini et al. "Infection-enhancing anti-SARS-CoV-2 antibodies recognize both the original Wuhan/D614G strain and Delta variants. A potential risk for mass vaccination?"[215]

The Path Forward

What is the path forward? What are the suggested steps required to end this now and make sure nothing like this happens again? How do we emerge?

1. **Never again should we use a "one-size-fits-all" approach**; instead, we should encourage an age-risk stratified "focused" protection approach, targeting only those who are at risk. Leave the rest of society alone, and definitely leave our healthy children alone who can effectively handle the virus (and most, if not all, pathogens) immunologically, especially with their potent (developing) innate immune system (innate antibodies and natural killer cells).

2. We need to **ensure to double and triple down protections of the elderly high-risk and vulnerable persons** in society (those with underlying medical conditions) in nursing homes, long-term care facilities, assisted-living facilities, care homes, in private households, etc.

3. **Allow physicians to exercise their best clinical judgements** in how they can best treat their patients without the threat of discipline and punitive actions for not following the approved political line on matters of natural immunity and vaccine safety. Medical license boards (state boards as well as the Colleges of Physicians and Surgeons in Commonwealth nations, e.g., Canada, United Kingdom, Australia, New Zealand, etc.) around the country and the world have threatened countless medical providers with punitive actions for informing patients as well as treating them early. The doctor-patient relationship used to be sacrosanct but that has been taken away. This has resulted in a neglect of early sequenced multi-drug treatment (combinations of antivirals, corticosteroids, and anti-thrombotic, anti-clotting drugs). There must be no ex cathedra over-reach by technocrats or bureaucrats in terms of how a doctor is to treat his or her patient.

4. There must be **urgent public service announcements** on vitamin D supplementation, on reducing obesity (maintaining a healthy body weight), as well as on the positive impact of the reduced risk to pathology due to healthy lifestyles, nutrition, exercise, etc.

5. **There must be a message to the population that we are not all at equal risk of severe outcomes** or death if infected (we never were), such that there is a one-thousand-fold difference in risk between children and older adults: ten-year-old Johnny who is in good health is not at the same risk of illness if exposed or infected as an eighty-five-year-old grandma who has two to three medical conditions.

6. There must be **no mass testing of asymptomatic persons**, only testing of symptomatic, ill/sick persons, including where there is a strong clinical suspicion. With this, stop contact tracing where the virus has already spread extensively as it confers no benefit; this tracing has actually been harmful.

7. There must be **no isolation/quarantine of asymptomatic persons**, only isolation of symptomatic ill/sick persons, including where there is a strong clinical suspicion. No isolation of asymptomatic persons at borders; this has been very harmful.

8. There must be **no mask mandates**, no mask use in school children, no mask use outdoors (it is nonsensical), and use of a mask is to be made on a case-by-case basis based on risks of the prevailing pathogen (the prevailing epidemiology).

9. There should be **no school closures,** no university closures, **nor forced quarantine** of people in contact with those who test positive for the virus.

10. There must be **no lockdowns whatsoever,** no business closures, and we must move to keep society open fully and immediately. The crushing harms and devastation from lockdowns, as we have seen, far outweigh any benefits and the harms are most pronounced among the poorer in society who are least able to afford the restrictions.

   The lockdown itself kills people, destroys families, and prevents education of our children. Child abuse was missed by closed schools (and remote schools) and the lockdowns promoted child abuse. Lost jobs caused stress in the household. There is near zero risk to children from COVID, and we are harming (and did harm) them by school closures; it was one of the most devastating misapplications of public policy. Most of the decisions made by the governments and their medical advisors were irrational, specious, and for the most part reckless, causing far greater harm. When schools were closed in America, millions of children went with no food as they get their only meals in the school setting.

   Countries like Canada, Australia, New Zealand, and Trinidad and Tobago of the Caribbean are test case examples of all that goes wrong with the nonsensical government-led responses and policies with unqualified, illogical, and irrational COVID advisors, ministries of health officials and leaders, medical officers of health, and a corrupted media running interference.

   The leaders of these nations should be fired from office for exacting an unbearable toll on their citizens through the ineptitude of their uninformed, irrational, and near-dictatorial actions, which had no scientific basis. This has played out now fully by the massive, accumulated data. They devastated their people and left them in a state of constant lockdown and reopening with no end in sight. They are incompetent because they failed to read the science or understand the lockdown data or evidence over two years, which is that it does not work in any manner and resulted in the mass suffering of the people. Business owners laid off employees, and school children could not take the anxiety, losses, and depression and committed suicide due to the lockdowns.

11.   **Always allow** (in such situations unless we are dealing with a
      pathogen with an elevated risk of death, etc.) **the vast majority
      of society**, the healthy, the young, and those with no underlying
      illnesses, **to continue normal daily life with reasonable
      common-sense precautions.** In other words, we do not impede
      the low risk of becoming infected and we leave them largely
      unrestricted with common sense safety precautions. We heighten
      their risk of transmission, increasing the probability of infection
      among the younger and low-risk persons, especially our healthy
      children. At the same time, we secure the high-risk-of-illness
      persons so that infection risk is reduced for them. This ensures
      that natural immunity emerges and we get closer to population-
      level herd immunity. We strongly mitigate the chance of infection
      in the high-risk. We create a risk differential of contracting the
      virus that is skewed toward the young and healthy. And we do this
      harmlessly and naturally.

12.   **Mandatory vaccination by a nation or setting is and was a
      non-starter;** there is no place for such mandates in societies that
      are free. No vaccinations for persons under seventy to seventy-five
      years of age (it is not needed and contra-indicated once there is
      no risk); COVID injections are to be offered, never mandated; no
      vaccinations for children[216] as the vaccine offers no opportunity
      for benefit[217] and only an opportunity for potential harms; no
      vaccination of pregnant women or women of child-bearing age;
      no vaccination of COVID-recovered persons (they have already
      cleared the virus and are now immune) or suspected COVID
      recovered persons. We never ever inject a biologically active agent
      into a pregnant woman, originally evidenced by the FDA that,
      knowing this, precluded them from the registrational trials.
      However, the same FDA then allowed pregnant women to be
      vaccinated. This was a catastrophic failure, and the implications
      shall emerge in time.

      If vaccines are used in persons over seventy to seventy-five
      years old, as suggested, they must only be used after shared
      decision-making with their clinicians, whereby patients can make
      informed decisions and provide informed consent.

13.   **Those advocating for vaccinations must also have risks on
      the table.** Thus, pharmaceutical companies, vaccine developers,

and governments, along with the FDA, must remove the liability protections.

No liability equates to no trust from the public, and certainly not from parents. Manufacturers must come to the table and, if they stand by these vaccines in that they are safe, then they (all involved in the manufacture and the advocating and mandating of these vaccines) must remove the liability protections that they benefit from. They must have direct skin in the game and be liable if there are harms as a result of the vaccinations.

14. **No vaccine passports** (or immunity or antibody passports) must ever be developed in these circumstances, as they are a violation of privacy, liberties, and personal freedoms. Such mandates constrain the rights of citizens under the questionable guise of safety; the vaccines as designed did not and do not protect an individual by the provision of "sterilizing immunity." By sterilizing immunity, we mean that there are neutralizing antibodies and there is no further prospect of either getting infected by the SARS-CoV-2 virus after a vaccination nor of passing along the virus to others. The evidence is and was very clear that the vaccines do no such thing and have failed, especially against the omicron variant, prompting even the CDC to state that the vaccinated and unvaccinated carry near equal virus and can spread alike.

A seminal and transformational Israeli study[218] by Gazit et al. clued us in and revealed that natural immunity conferred longer-lasting and stronger protection against infection, symptomatic disease, and hospitalization caused by the delta variant of SARS-CoV-2, compared to the BNT162b2 two-dose vaccine-induced immunity. Vaccinees who had not been previously exposed to the virus had a 13.06-fold (95 percent confidence index [CI], 8.08 to 21.11) increased risk for breakthrough infection with the delta variant compared to those previously infected. A massive trove of evidence has emerged to show that the vaccinated and unvaccinated carry the same risk of infection and potential transmission.[219]

15. **The FDA and the CDC with vaccine developers must immediately implement proper safety surveillance systems** for these COVID (and any subsequent) vaccines. This must include data safety monitoring boards post-vaccination, critical event

committees, and ethics review committees, which, at this time (a year and a half after vaccine rollout), still do not exist. With this, a committee to review the existence and proper administration of ethical and fully informed consent by vaccine candidates should be established.

16. Make clear that **a "case" is when someone has symptoms and is sick;** an "infection" is not a "case," and this effort to deceive the public with the reporting of "cases" must stop immediately so that the public understands the accurate parameters of the emergency.

17. **Implement immediate testing for antibody and T-cell immunity before vaccinating** the designated group. If we are vaccinating the higher-risk persons, we do not vaccinate persons who have active infection or who have recovered from infection, the same way if your child gets the measles infection and displays the rash and fever, etc., you do not then vaccinate them after they have recovered. You send them to school for they are now immune; use that same logic with COVID-19.

18. **Cease the** illogical, irrational, inaccurate, and nonsensical **absurdity that COVID-19 vaccine immunity is superior to naturally acquired immunity** when the science is clear that natural exposure immunity is broad, robust, durable, mature, long-lasting, and similar to, if not way superior to, the narrow and immature immunity conferred by the COVID vaccines. An article by Scott Morefield of the Brownstone Institute titled "The Infuriating Habit of Ignoring Natural Immunity,"[220] reveals the ridiculousness of the CDC and NIH on this point.

　　Just look at this article from Israel on research looking at infection from previously infected individuals and recovered double vaccinated individuals.[221] It essentially destroys the negation of natural immunity or need for vaccination or vaccine passports *in toto*.

> More than 7,700 new cases of the virus have been detected during the most recent wave starting in May, but just 72 of the confirmed cases were reported in people who were known to have been infected previously—that is, less than 1% of the new cases. Roughly 40% of new cases—or more than 3,000 patients— involved people who had been infected despite being vaccinated. With a total of 835,792 Israelis known to have recovered from the

virus, the 72 instances of reinfection amount to 0.0086% of people who were already infected with COVID. By contrast, Israelis
who were vaccinated were 6.72 times more likely to get infected
after the shot than after natural infection, with over 3,000 of the
5,193,499, or 0.0578%, of Israelis who were vaccinated getting
infected in the latest wave.

Moreover, this review by me of 150 studies and published by the
Brownstone Institute on natural immunity shows the robust and
complete nature of natural immunity.[222]

19. It is way past time to **throw away the masks for our children** as
they have provided no benefit, yet can cause harm to the growing
child emotionally, socially, and in their health and well-being. The
masks are toxic, especially to our children.[223] Unshackle your
children, allow them to play freely outside with their friends and
to breathe the fresh air; allow your children again to live naturally
within their environments. Allow their immune systems (their
natural innate immunity system, their mucosal immunity) to be
taxed and tuned up daily, challenged by the outdoors, by mingling
and socially interacting, by living as normally as possible.

We are potentially creating a disaster and have likely set our
children up for immunological disaster by the lockdowns, the
masking, and school closures that have weakened their developing
immune systems. Moreover, the COVID injections can potentially
devastate their developing innate immune system (subvert and
sideline it).

20. **The public must shift toward the use of nasal-oral hygiene
washes** that have been shown to be highly effective in not just
eliminating the COVID virus but also a range of pathogens.
These include povidone-iodine 10% solution[224] (can be purchased
over-the-counter) and diluted as well as hydrogen peroxide when
PI is not palatable. These must be diluted and never swallowed
but used to clean out the oral and nasal passages of pathogens,
especially if there was a high risk of exposure (i.e., you have been
in crowded settings). These are so very effective that they can help
minimize even the use of sequenced multi-drug early treatment[225]
(combinations of anti-viral, corticosteroid, and anti-platelet
therapeutics), and we have now inserted it into the second row just
beneath self-quarantine at home in our advisory guidelines.

Conclusion

In closing, the medical experts and the COVID task forces have been wrong. Every decision in the United States, Canada, the United Kingdom, etc. has proven disastrous, and they have caused far greater suffering and death from the collateral effects of the lockdowns and restrictions. The medical experts who informed governments should have broadened the scope of advice and allowed other voices to be heard. They failed to do so and, in return, caused crushing harms by inept and unsound decisions.

If it is all about the science, medical decision-makers must follow the data and science and use it, as well as use critical analysis of the data. These decision-makers must now come to understand the catastrophic impact of their policies and that stopping COVID at all costs (so-called "zero COVID") is not a policy and it is not—and certainly was never—attainable. If a policy is based on an unattainable goal, pursuing it by every mean necessary causes greater harm to the population. Our decision-makers failed to understand this and this resulted in serious harms and needless deaths in our populations, beyond what any virus did.

The path forward begins with immediately obliterating COVID restrictions that include all mandates, all mass vaccination strategies, and a move to re-establish as normal a life as possible. We can begin by considering these twenty points and adding, as well as implementing, those that you think could be beneficial to making society as whole as possible as soon as possible.

EPILOGUE

What the Data Show Now

An article from September 2022 from the *New York Times* detailed the harm done by the lockdown on American schoolchildren:

> This year, for the first time since the National Assessment of Educational Progress tests began tracking student achievement in the 1970s, 9-year-olds lost ground in math, and scores in reading fell by the largest margin in more than 30 years.
>
> The declines spanned almost all races and income levels and were markedly worse for the lowest performing students. While top performers in the 90th percentile showed a modest drop—three points in math—students in the bottom 10th percentile dropped by 12 points in math, four times the impact.
>
> "I was taken aback by the scope and magnitude of the decline," said Peggy G. Carr, Commissioner of the National Center for Education Statistics, the federal agency that administered the exam earlier this year. The tests were given to a national sample of 14,8000 9-year-olds and were compared with the results of tests taken by the same age group in early 2020, just before the pandemic took hold in the United States.[226]

The loss of progress is American schoolchildren is undeniable, even to the *New York Times*.

And all of it was unnecessary.

Because of the fact that the ACE-2 receptor, which allows entry of SARS-CoV-2 into the cell, is not generally active in children or expressed in greater numbers in the nasal-oral passages, they were at essentially zero risk. Because they come with a primed, pre-activated innate immune system ready to do battle with COVID and a broad range of pathogen (viruses) as it further gets trained and matures. Because children show elevated cross-reactivity (and potential cross-protection) to a range of coronaviruses due to prior exposure to common colds . Because the risk from day one in February 2020, to now, September 2022, has remained statistically zero in children to becoming severely ill or dying from COVID.

What can we say about the teacher's unions, which were screaming that they needed to protect the teachers from their students?

They were wrong. Flat wrong and they even had the data to show they were lying.

Teachers should have been standing up in their classrooms, mask-less, looking at the smiling faces of their mask-less students. It was the teacher who presented risk to the children, not the other way around, and the data was bulletproof on that. The median/mean age of American teachers is approximately forty-one years old, and as such they have been at low risk throughout. They are generally healthy persons.

In an abundance of caution, teachers should have only been required to wear masks when they were around other teachers and, if they wanted a remote or hybrid model for themselves case-by-case, based on their risk, then so be it. Children should have never been made to pay the price by the CDC and the unions.

But these decisions by public health authorities in the United States certainly made a lot of parents unhappy.

And they took out that anger on President Donald Trump.

* * *

How much should you trust an "internal review" by any organization?

Not much, in my opinion.

However, in the wake of the COVID-19 crisis, the CDC did perform such an internal audit of their operation and the director spoke publicly. Did they find all the problems, or simply highlight the ones that were undeniable? This is how it was framed in August 2022 by the *Washington Post*:

> The nation's top public health official acknowledged Wednesday that the Centers for Disease Control and Prevention had failed to respond effectively to the coronavirus pandemic, and announced plans for extensive changes,

including faster release of scientific findings and easier-to-understand guidance.

CDC Director Rochelle Walensky told senior leaders she was committed to a long-sought revamp of CDC culture based on an internal review that called for a more nimble and better-trained workforce and incentives that would reward actions over publication – moves that allies said were necessary and critics said could not come quickly enough.[227]

Are we to genuinely believe that the problem was simply that they took too long to get the information out and they needed to use simpler language?

They need to be "more nimble"?

What the hell does that mean?

Ballet in the halls?

Yoga?

Speed training?

They need to be "better trained"?

They've supposedly got master's in Public Health, MDs, and PhDs.

Maybe they're just idiots. Maybe they are purely inept and do not read the data or science, do not "get" it, or simply cannot understand the science. The CDC as the marquee public health agency in the United States has shown a remarkable depth of cognitive dissonance to any information disparate from its narrative, and a shocking depth of academic sloppiness.

Maybe they just come up with stupid ideas.

Maybe they don't know how to assess evidence.

Did they investigate these questions?

I think not.

But they had to at least give the appearance of doing an evenhanded job, didn't they? The article continued:

Rich Besser, president and CEO of the Robert Wood Johnson Foundation, a health philanthropy group, said he was interviewed as part of CDC's internal review and agreed with Walensky's ambitions for the agency. But he also cautioned that the plan required cooperation from Congress, more transparency from CDC itself—and a realistic assessment of the challenges of overhauling a federal workforce that has been scattered during the pandemic.

"I worry about the ability to affect culture change, when the agency is still largely remote," said Besser, who served as acting CDC director during the Obama administration. "I would want to be standing in front of the agency and laying out a vision and inspiring people toward that change. And

then I would be walking the halls . . . but my understanding is that the build-
ings are still pretty much empty."[228]

Can somebody please explain to me what is going on in what is supposedly the world's greatest public health agency? These medical and scientific
experts can't be expected to do their job without somebody walking up and
down the halls, constantly acting like a high school football coach, and
telling them they can win the big game and date the homecoming queen?

And let me ask you, do you consider a former "acting director of the
CDC" to be a genuine critic?

There needs to be an independent review of the CDC.

And if the problems are as bad as believed by the two senators and congressman who wanted me to write that report, then I understand why they
wanted to demolish the CDC down to its studs.

And maybe along the way we can throw a few people in jail, like they
did after the 2008 home mortgage meltdown.

Oh yeah, none of those financial guys were ever convicted of a single
thing.

And the band plays on as big money names the tune.

* * *

I believe it's important that Donald Trump (assuming his health remains
good), or somebody with his populist principles, run for president in 2024.

This is my personal argument for why Trump, despite the mistakes he
has made, would be the best candidate.

I believe his years as a real estate guy have given him a unique perspective. When he chose to run in 2016, I think he looked at the US government
as a building that needed a remodel. It was a glorious building, with a hal-
lowed history, but it was run down.

But like many people who buy a property, only to learn how many
things actually need to be fixed, his perspective changed once he got the
keys to the place and moved in.

The government needed more than a remodel.

It requires a tear-down and rebuild.

As I watch Trump today, endorsing candidates in Republican prima-
ries, and building up an enormous winning streak, I see he is gathering the
team around him that he needs for this historic endeavor.

In his first term, he was surrounded by people who did not share his vision or agenda.

In his second term, it needs to be much different.

The Democrats and the bureaucratic Deep State see things exactly as I do, which is why they are trying so desperately to knock him out of the running in 2024. They sense that a second term President Trump will properly finish the job he started.

In their eyes they must not let this happen.

I know there are those who will criticize Trump for allowing the lockdowns to take place (as I too have been critical), as well as the enormous amount of pride he seems to currently take in Operation Warp Speed, which, as I've said, I believe will go down as the worst public health disaster in history.

These were mistakes Trump made because he did not follow his instincts. He also trusted the counsel he was getting. He is human.

Early in his term, Trump said he was going to appoint Robert F. Kennedy Jr. to lead a committee to look at the safety of vaccines in general. You have to understand my view on Kennedy. I think he would have been a great advocate for the safety of vaccines, and he would have helped protect our children. I find him brilliant and not given the credit he deserves. I have met him and interviewed with him and find him charismatic and intense, well read and very well prepared to talk. He knew of studies and papers I have written and published, and I was shocked that he read all of my science before we talked, down to the minutiae and very technical issues. He would have looked after the best interest of the American public and maybe we may not have been in the devastating mess we are in now with these ineffective and unsafe COVID gene injections. In the meeting Trump had with Kennedy, Trump personally told him of several friends of his who had a normally developing child, who, after a series of childhood vaccinations, developed autism.

Their stories were all the same, and it was clear Trump had been deeply touched by their accounts.

Since my time in government, I have come to know Robert F. Kennedy Jr. and believe him to be one of the most honest and important figures of our time, which is why the pharma controlled social media giants are using all of their tricks so that the American people do not hear his voice.

However, that effort was scuttled, perhaps by Dr. Anthony Fauci, or even by Bill Gates, who publicly took credit for stopping the committee.[229]

Would there have ever been a COVID-19 crisis, at least to the extent it became, if Trump had allowed Robert F. Kennedy Jr.'s vaccine safety committee to go forward?

I think not, and that is a terrible tragedy.

From my time in government and what I have observed as a private citizen, Trump is one of the few leaders who considers himself to be in a genuine relationship with the public. It is almost like an ongoing love affair he is having with the public, where each gives something to the other. I believe that's why he loves doing his rallies so much—because it gives him a way to gauge the sentiment of the public. He wants to do well by the American people.

That's why books like this one are so important, because it's how average people can see through the lies of big pharma and big science.

And in turn, you can convince Trump to change his mind.

Trump knows the bureaucratic Deep State hates him.

Trump knows the military-industrial complex hates him.

Trump knows the big tech giants hate him.

Trump knows the federal agencies that hate him.

But Trump also knows he has a special bond with the public who trust him.

And when you are in a relationship, you listen to what the other person says, without worrying about what you are going to say in return.

Trump has this ability to a greater degree than any other politician today. We must all do our part. I continue to write and speak out. In Canada, I participated in the trucker convoy which put the fear of God in Justin Trudeau and nearly got myself arrested. I am not an outlaw. I am simply a patriot, raising my voice, until the leaders answer our questions. In the United States, I was also part of the lockdown protests. I have no choice but to speak against these immoral policies. If you have read this book, I think you also probably feel a similar sense of obligation to join the fight.

Together, I am confident we can fix these problems, and bring about an unprecedented era of good health and prosperity for all.

Endnotes

Introduction

1. Patrick Svitek, "In Texas, President Trump's Team Hits the Road to Shore Up Support," *Texas Tribune*, September 3, 2020, www.texastribune.org/2020/09/03/texas-president-trump-bus-tour/.
2. Eric Schumaker, "Anthony Fauci Says One Line from 'The Godfather' Has Shaped His Career," *Business Insider*, October 7, 2021, www.businessinsider.com/anthony-fauci-line-from-the-godfather-shaped-career-2021-10.
3. Mike Lauer, "FY 2020 by the Numbers: Extramural Investments in Research," *National Institutes of Health: Office of Extramural Research*, April 21, 2021, nexus.od.nih.gov/all/2021/04/21/fy-2020-by-the-numbers-extramural-investments-in-research/.
4. Karl Evers-Hillstrom, "Most Expensive Ever: 2020 Election Cost $14.4 Billion," *Open Secrets*, February 11, 2021, www.opensecrets.org/news/2021/02/2020-cycle-cost-14p4-billion-doubling-16/.
5. Editorial Board, "How Fauci and Collins Shut Down COVID Debate," *Wall Street Journal*, December 22, 2021, www.wsj.com/articles/fauci-collins-emails-great-barrington-declaration-covid-pandemic-lockdown-11640129116?mod=Searchresults_pos1&page=1.
6. Ibid.
7. Fred Guterl, "Dr. Fauci Backed Controversial Wuhan Lab with U.S. Dollars for Risky Coronavirus Research," *Newsweek*, April 28, 2020, www.newsweek.com/dr-fauci-backed-controversial-wuhan-lab-millions-us-dollars-risky-coronavirus-research-1500741.
8. Audrey McNamara, "Former CDC Chief Says 'Most Likely' Cause of Coronavirus is that it 'Escaped" from a Lab," *CBS News*, March 27, 2021, https://www.cbsnews.com/news/covid-lab-theory-robert-redfield-no-evidence/.
9. Ethan Ennals, "Trump Aide Claims Covid 'Came out of the Box Ready to Infect' – Claiming Virus was Being Worked on by Scientists in a Chinese Lab," *Daily Mail*, July 16, 2022, www.dailymail.co.uk/news/article-11021329/Trump-aide-claims-Covid-came-box-ready-infect.html.
10. "Coronavirus Death Toll and Trends," *Worldmeter*, (Accessed August 6, 2022), www.worldometers.info/coronavirus/coronavirus-death-toll/.

11   Sharon Lerner, "Jeffrey Sachs Presents Evidence of Possible Lab Origin of COVID-19," *The Intercept*, May 19, 2022, www.theintercept.com/2022/05/19/covid-lab-leak -evidence-jeffrey-sachs/.

12   Nathan Robinson, "Why the Chair of Lancet's COVID-19 Commission Thinks the US Government is Preventing a Real Investigation in the Pandemic, *Current Affairs*, August 2, 2022, www.currentaffairs.org/2022/08/why-the-chair-of-the-lancets-covid -19-commission-thinks-the-us-government-is-preventing-a-real-investigation-into-the -pandemic.

13   Ibid.

14   Office of the Chief of Staff, National Institute for Allergy and Infectious Diseases, (Accessed August 2, 2022), www.niaid.nih.gov/about/chief-staff-contacts.

15   Email from Woleola Akinso to Members of the COVID-19 Task Force, September 8, 2020, 1:04 p.m.

16   Email from Dr. Paul Alexander to COVID-19 Task Force, September 8, 2020, 1:35 pm.

17   Paul Elias Alexander, "75 Studies and Articles Against COVID-19 School Closures," December 24, 2021, *Brownstone Institute*, www.brownstone.org/articles/75-studies-and -articles-against-covid-19-school-closures/.

18   Sarah D. Marks, "Child Abuse Cases Got More Severe During COVID-10. Could Teachers Have Stopped It?" *Education Week*, June 1, 2021, www.edweek.org/leadership /child-abuse-cases-got-more-severe-during-covid-19-could-teachers-have-prevented-it /2021/06.

19   Alicia Betz, "What it Means that Teachers are Mandated Reporters," 2022, *Education Corner*, www.educationcorner.com/teachers-mandated-reporters.html.

20   Email from Jennifer Routh to COVID-19 Task Force, September 8, 2020, 2:44 pm.

21   Email from Dr. Paul Alexander to COVID-19 Task Force, September 8, 3:31 pm.

22   Email from Dr. Andrea Lerner to COVID-19 Task Force, September 8, 2020, 5:07 pm.

23   Letter from Dr. Paul Alexander to COVID-19 Task Force, September 8, 2020, 5:07 pm.

24   Kosta Danis, Oliver Epaulard, Thomas Bent, et al, "Cluster of Coronavirus Disease (COVID-19) in the French Alps, *Journal of Clinical Infectious Diseases*, February 2020, August 1, 2020, www.ncbi.nlm.nih.gov/pmc/articles/PMC7184384/.

25   Cody B. Jackson, Michael Farzan, Bing Chen, & Hyerun Choe, "Mechanisms of SARS-CoV-2 Entry into Cells," *Nature Reviews Molecular Cell Biology*, October 5, 2021, www .nature.com/articles/s41580-021-00418-x.

26   Supinda Bunyavanich, Anh Do, and Alfin Vicenio, "Nasal Gene Expression of Angiotensin-Converting Enzyme in Children and Adults," *Journal of the American Medical Association*, June 16, 2020, Vol. 323, Number 23, www.pubmed.ncbi.nlm.nih .gov/32432657/.

27   Ibid.

Chapter One

28   Sascha Wilson, "Sando Mayor Unveils Dr. Eric Williams Plaque – Ex-PM Must Be on School Curriculum," *Trinidad and Tobago Guardian*, August 19, 2013, www.guardian. co.tt/article-6.2.406368.169454e6e9.

29   Donald G. McNeil Jr., "D.A. Henderson, Doctor Who Helped End Smallpox Scourge, Dies at 87," *New York Times*, August 21, 2016, www.nytimes.com/2016/08/22/us /dr-donald-a-henderson-who-helped-end-smallpox-dies-at-87.html#:~:text=He%20 was%2087.,said%20his%20daughter%2C%20Leigh%20Henderson.

30   Roger L. Sur and Phillipp Dahm, "History of Evidence-Based Medicine," *Indian Journal of Urology*, Oct-Dec: 27(4): 487-489; doi: 10.4103/0970-1591.91438.

31   "History of Evidence-Based Practice," Clinical Information Access Portal, (Accessed October 15, 2022), www.ciap.health.nsw.gov.au/training/ebp-learning-modules/module1/history-of -evidence-based-practice.html.

Chapter Two

32   Sandra Gonzalez, "Brad Pitt Scores Emmy Nomination for Playing Dr. Fauci on SNL," *CNN*, July 29, 2020, www.cnn.com/2020/07/28/entertainment/brad-pitt-emmy -nomination-fauci/index.html.

33   Etan Ennals, "Trump Aide Claims Covid 'Came out of the Box Ready to Infect' – Claiming Virus was Being Worked on by Scientists in a Chinese Lab," *Daily Mail*, July 16, 2022*l*, www.dailymail.co.uk/news/article-11021329/Trump-aide-claims-Covid-came -box-ready-infect.html.

34   Jessie Yeung, "The Risk of a Global Pandemic is Growing – And the World Isn't Ready, Experts Say," *CNN*, September 18, 2019, www.cnn.com/2019/09/18/health/who -pandemic-report-intl-hnk scli.

35   Ibid.

36   "Reconstruction of the 1918 Influenza Pandemic Virus," CDC Website, (Last reviewed: December 17, 2019), www.cdc.gov/flu/about/qa/1918flupandemic.htm.

37   Ibid.

38   Ibid.

39   "A World at Risk – Annual Report on Global Preparedness for Health Emergencies," September 2019, *Global Preparedness Monitoring Board*, www.gpmb.org/annual-reports /annual-report-2019.

40   Ibid.

41   Ibid.

42   Ibid.

43   Ibid.

44   Ibid.

45   Ibid.

46   Ibid.

47   Emily Kirkpatrick, "Melinda Gates Says Bill Gates's Work with 'Abhorrent' Jeffrey Epstein Led to Divorce," *Vanity Fair*, March 3, 2022, www.vanityfair.com/style/2022/03 /melinda-gates-jeffrey-epstein-led-to-bill-gates-divorce-gayle-king-interview.

48   Thomas V. Inglesby, Jennifer B. Nuzzo, Tara O'Toole, and D.A. Henderson, "Disease Mitigation in the Control of Pandemic Influenza," *Biosecurity and Bioterrorism: Biodefense Strategy, Practice, and Science*, Vol. 4, Number 4, 2006, p. 366-375, www .pubmed.ncbi.nlm.nih.gov/17238820/, doi: 10.10.1089/bsp.2006.4.366.

49   Ibid. at p. 368-369.

50   Ibid. at p. 373.

51   Ibid.

52 "Non-Pharmaceutical Public Health Measures for Mitigating the Risk and Impact of Epidemic and Pandemic Influenza," *World Health Organization*, 2019, www.apps.who.int/iris/bitstream/handle/10665/329438/9789241516839-eng.pdf.

53 Ibid. at p. 13-18.

54 Ibid. at p. 3.

55 Ibid. at p. 2.

56 Jonas Harby, Lars Jonung, and Steve H. Hanke, "A Literature Review and Meta-Analysis of the Effects of Lockdowns on COVID-19 Mortality," *Studies in Applied Economics*, No. 200, January 2002, www.sites.krieger.jhu.edu/iae/files/2022/01/A-Literature-Review-and-Meta-Analysis-of-the-Effects-of-Lockdowns-on-COVID-19-Mortality.pdf.

57 Ibid. at p. 29.

58 "Newly Released Estimates Show Traffic Fatalities Reached a 16 Year High in 2021," *National Highway Traffic Safety Administration*, May 17, 2022, www.nhtsa.gov/press-releases/early-estimate-2021-traffic-fatalities#:~:text=NHTSA%20projects%20that%20an%20estimated,Fatality%20Analysis%20Reporting%20System's%20history.

59 Jonas Harby, Lars Jonung, and Steve H. Hanke, "A Literature Review and Meta-Analysis of the Effects of Lockdowns on COVID-19 Mortality," *Studies in Applied Economics*, No. 200, January 2002, p. 43, www.sites.krieger.jhu.edu/iae/files/2022/01/A-Literature-Review-and-Meta-Analysis-of-the-Effects-of-Lockdowns-on-COVID-19-Mortality.pdf.

60 John Tierny, "Fauci and Walensky Double Down on Failed Covid Response," *Wall Street Journal*, August 18, 2022, www.wsj.com/articles/fauci-and-walensky-double-down-on-failure-covid-pandemic-evidence-data-lockdowns-mask-mandates-restrictions-public-health-experts-11660855180.

61 Ibid.

62 Ibid.

63 Ibid.

64 Deborah Birx, *Silent Invasion: The Untold Story of the Trump Administration, Covid-19, and Preventing the Next Pandemic Before It's Too Late*, (New York, New York: Harper Collins, 2022), 250-251.

65 John Tierny, "Fauci and Walensky Double Down on Failed Covid Response," *Wall Street Journal*, August 18, 2022, www.wsj.com/articles/fauci-and-walensky-double-down-on-failure-covid-pandemic-evidence-data-lockdowns-mask-mandates-restrictions-public-health-experts-11660855180.

66 "CDC Director Orders Agency Overhaul, Admitting Flawed Covid-19 Response," *Politico*, August 17, 2022, www.politico.com/news/2022/08/17/cdc-agency-overhaul-covid-19-response-00052384.

67 Ibid.

68 Ibid.

69 John Tierny, "Fauci and Walensky Double Down on Failed Covid Response," *Wall Street Journal*, August 18, 2022, www.wsj.com/articles/fauci-and-walensky-double-down-on-failure-covid-pandemic-evidence-data-lockdowns-mask-mandates-restrictions-public-health-experts-11660855180.

70 Ibid.

Chapter Three

71 "Explaining Operation Warp Speed," NIH Website, (Accessed August 26, 2022), www
    .niaid.nih.gov/grants-contracts/what-operation-warp-speed.

72 Ibid.

73 Joseph Sonnabend, "The Long Road to PCP Prophylaxis in AIDS. An Early History,"
    Poz, September 23, 2009, www.poz.com/blog/the-long-road-to-PCP.

74 Ibid.

75 Ibid.

76 Ibid.

77 Robert F. Kennedy, Jr, *The Real Anthony Fauci: Bill Gates, Big Pharma, and the Global
    War on Democracy and Public Health*, (New York, New York, Skyhorse Publishers,
    2021), 151.

78 Ibid at p. 166.

79 Jade Scipioni, "The Supplements Dr. Fauci Takes to Help Keep His Immune System
    Healthy," *CNBC*, September 14, 2020, www.cnbc.com/2020/09/14/supplements
    -white-house-advisor-fauci-takes-every-day-to-help-keep-his-immune-system-healthy
    .html.

80 Ibid.

81 Senator Ron Johnson, "Vaccine Mandates," Congressional Record, Vol. 167, No. 212,
    (Senate – December 8, 2021), www.congress.gov/congressional-record/volume-167
    /issue-212/senate-section/article/S9002-2.

82 Adam Adrezejewski, "Dr. Anthony Fauci Received big Pay Increase to Prevent
    Pandemics," *Forbes*, October 20, 2021, www.forbes.com/sites/adamandrzejewski
    /2021/10/20/dr-anthony-faucis-little-known-biodefense-work--its-how-he-became-the
    -highest-paid-federal-employee/?sh=739dd6b16081.

83 Tara O'Neill Hayes & Serena Gillian, "Chronic Disease in the United States: A
    Worsening Health and Economic Crisis," *American Action Forum*, September 10,
    2020,   www.americanactionforum.org/research/chronic-disease-in-the-united-states-a
    -worsening-health-and-economic-crisis/.

84 Saron Lerner, Mara Hvistendahl, & Maia Hibbett, "NIH Documents Provide New
    Evidence U.S. Funded Gain of Function Research in Wuhan," *The Intercept*, September
    9, 2021, www.theintercept.com/2021/09/09/covid-origins-gain-of-function-research/.

85 Michael Lee, "Outgoing NIH Director says 'Hundreds of Thousands Would Have
    Died' From COVID if US Hadn't Listened to Him," *Fox News*, December 19, 2021,
    www.foxnews.com/politics/outgoing-nih-director-says-hundreds-of-thousands-would
    -have-died-from-covid-if-us-hadnt-listened-to-him.

86 Ibid.

87 Ibid.

88 Antonio Regalado, "No One Can Find the Animal that Gave People Covid-19," *MIT
    Technology Review*, March 26, 2021, www.technologyreview.com/2021/03/26/1021263
    /bat-covid-coronavirus-cause-origin-wuhan/.

89 Ibid.

90 Ibid.

91 Ibid.

92 Ibid.

93 "Bio of Jay Bhattacharya," *Stanford Profiles*, (Accessed August 24, 2022), www.profiles
    .stanford.edu/jay-bhattacharya.

94   "Sunetra Gupta," *Merton College, Oxford,* (Accessed August 24, 2022), www.merton.ox
     .ac.uk/professor-sunetra-gupta.
95   "Martin Kulldorff," *Hillsdale College,* (Accessed August 24, 2022).
96   Martin Kulldorff, Sunetra Gupta, and Jay Bhattacharya, "Great Barrington Declaration,"
     October 4, 2020, www.gbdeclaration.org.

Chapter Four

97   Daphne Khalida Kilea, "The Year in Review: Some of the 2019 Events that Rocked
     the Nation," *VC Reporter,* December 25, 2019, www.vcreporter.com/news/the-year-in
     -review-some-of-the-2019-events-that-rocked-the-nation/article_97a82178-18f3
     -5d31-b4e5-905927586d35.html.
98   Veronia Rocha, "Trump Feeds Fish, Winds Up Pouring Entire Box of Food into Koi
     Pond," *CNN,* November 6, 2017, www.cnn.com/2017/11/06/politics/donald-trump
     -koi-pond-japa.
99   Steve Cortes, "Trump's Top 10 Achievements for 2019," *Real Clear Politics,*
     December 31, 2019, www.realclearpolitics.com/articles/2019/12/31/trumps_top_10
     _achievements_for_2019_142047.html.
100 Ibid.
101 Ibid.
102 "Trump Administration Accomplishments – As of January 2021," White House
     Archives, www.trumpwhitehouse.archives.gov/trump-administration-accomplishments/.
103 Allison Aubrey, "Trump Declares Coronavirus a Public Health Emergency and Restricts
     Travel from China," *NPR,* January 31, 2020, https://www.npr.org/sections/health
     -shots/2020/01/31/801686524/trump-declares-coronavirus-a-public-health-emergency
     -and-restricts-travel-from-china.
104 Maggie Haberman and Jonathon Martin, "Trump's Re-election Chances Suddenly
     Look Shakier," *New York Times,* March 12, 2020, www.nytimes.com/2020/03/12/us
     /politics/trump-vs-biden.html.
105 Kai Kupterschmidt and Jon Cohen, "China's Aggressive Measures Have Slowed the
     Coronavirus. This May Not Work in Other Countries, *Science,* March 2, www.science
     .org/content/article/china-s-aggressive-measures-have-slowed-coronavirus-they-may
     -not-work-other-countries.
106 Ibid.
107 Peter Daszak, "We Knew Disease X was Coming. It's Here Now," *New York Times,*
     February 27, 2020, www.nytimes.com/2020/02/27/opinion/coronavirus-pandemics
     .html?.
108 Birx, 129.
109 Ibid at p. 147.
110 Ibid. at pp. 150-151.
111 Ibid. at p. 152.
112 Michael Senger, "The Talented Mr. Pottinger: The U.S. Intelligence Agent Who
     Pushed Lockdowns," *Brownstone Institute,* July 20, 2022, www.brownstone.org/articles
     /the-talented-mr-pottinger-the-us-intelligence-agent-who-pushed-lockdowns/.
113 Ibid.

[114] Yasmeen Abutaleb and Damian Paletta, *Nightmare Scenario: Inside the Trump Administration's Response to the Pandemic that Changed History* (New York, New York: Harper Collins, 2021).

[115] "Yen Pottinger – Distinguished Alumnae," Emma Willard School, (Accessed September 3, 2022), www.emmawillard.org/list-detail?pk=195687&fromId=279876.

[116] Birx, 32-33.

[117] Ibid. at p. 38.

[118] Ibid. at p. 62.

[119] Michael Senger, "The Talented Mr. Pottinger: The U.S. Intelligence Agent Who Pushed Lockdowns," *Brownstone Institute*, July 20, 2022, www.brownstone.org/articles/the-talented-mr-pottinger-the-us-intelligence-agent-who-pushed-lockdowns/.

[120] Mark Meadows, *The Chief's Chief*, (Saint Petersburg, Florida, All Seasons Press, November 2021), 50-51

[121] Ibid. at p. 51.

[122] Ibid. at p. 54.

[123] Ibid.

[124] Ibid. at p. 57.

[125] World Health Organization, "January 14, 2020, 3:18 am), *Twitter*, www.twitter.com/who/status/1217043229427761152.

[126] Meadows, 55.

[127] Ibid. at pp. 55-56

[128] Niraj Chokshi, "That Wasn't Mark Twain: How a Misquotation Is Born," *New York Times*, April 26, 2017, https://www.nytimes.com/2017/04/26/books/famous-misquotations.html.

[129] "Michael Crichton Quotes," *GoodReads*, (Accessed October 15, 2022), www.goodreads.com/quotes/344539-i-want-to-pause-here-and-talk-about-this-notion

[130] Scott W. Atlas, *A Plague Upon Our House: My Fight at the Trump White House to Stop COVID from Destroying America*, (New York, New York, Post Hill Press, 2021), 117-118.

[131] Ibid. at p. 199.

Chapter Five

[132] Pfizer Press Release "Pfizer and BioNTech Announces Vaccine Candidate Against COVID-19 Achieved Success in First Interim Analysis from Phase 3 Study," Pfizer website, November 9, 2020, www.pfizer.com/news/press-release/press-release-detail/pfizer-and-biontech-announce-vaccine-candidate-against.

[133] Jon Cohen, "Fact Check: No Evidence Supports Trump's Claim that COVID-19 Vaccine Result was Suppressed to Sway Election", *Science*, November 9, 2020, www.science.org/content/article/fact-check-no-evidence-supports-trump-s-claim-covid-19-vaccine-result-was-suppressed.

[134] Ibid.

[135] Gretchen Vogel, "Omicron Booster Shots are Coming – With Lots of Questions," *Science Insider*, August 30, 2022, www.science.org/content/article/omicron-booster-shots-are-coming-lots-questions.

[136] VAERS COVID Vaccine Adverse Events Reports," Open VAERS, (Accessed September 6, 2022), www.openvaers.com/covid-data.

[137] Ibid.
[138] Ibid.
[139] Ibid.
[140] Ibid.
[141] Ibid.
[142] Ibid.
[143] Ibid.
[144] Ibid.
[145] Ibid.
[146] Ibid.
[147] Ibid.
[148] Ibid.
[149] Ibid.
[150] Ibid.
[151] Ibid.
[152] Ibid.
[153] Ibid.
[154] Ibid.
[155] Ibid.
[156] Ibid.
[157] Ibid.
[158] David Kessler, "Introducing MedWatch: A New Approach to Reporting Medication and Device Adverse Effects and Product Problems," *Journal of the American Medical Association*, June 21, 1993, Vol. 269, No. 21, www.fda.gov/downloads/Safety /MedWatch/UCM201419.pdf.
[159] Spiro Pantazatos and Herve Seligmann, "COVID-Vaccination and Age-Stratified All-Cause Mortality Risk," *Research Gate*, October 2021, www.researchgate.net /publication/355581860_COVID_vaccination_and_age-stratified_all-cause_mortality _risk.
[160] Ibid.

Chapter Six

[161] Yasmeen Abutaleb, Lena H. Sun, and Rosalind S. Helderman, "Top Trump Health Appointee Michael Caputo Warns of Armed Insurrection after Election," *Washington Post*, September 14, 2020, www.washingtonpost.com/health/2020/09/14/michael -caputo-coronavirus-cdc/.
[162] Ibid.
[163] Charon LaFraniere, "Trump Health Aide Pushes Bizarre Conspiracies and Warns of Armed Revolt," *New York Times*, September 14, 2020, www.nytimes.com/2020/09/14 /us/politics/caputo-virus.html.
[164] Ibid.
[165] Noah Manskar, "Riots Following George Floyd's Death May Cost Insurance Companies up to $2 Billion," *New York Post*, September 16, 2020, www.nypost.com/2020/09/16 /riots-following-george-floyds-death-could-cost-up-to-2b/.
[166] "How Many Guns are in the U.S.?" American Gun Facts, (Accessed September 7, 2022), www.americangunfacts.com/gun-ownership-statistics/.

167 Charon LaFraniere, "Trump Health Aide Pushes Bizarre Conspiracies and Warns of Armed Revolt," *New York Times*, September 14, 2020, www.nytimes.com/2020/09/14/us/politics/caputo-virus.html.

168 Ibid.

169 Ibid.

170 Chas Danner, "A Guide to the Intense Debate over Biden's Big Democracy Speech," *New York* magazine September 6, 2022, www.nymag.com/intelligencer/2022/09/a-guide-to-the-intense-debate-over-bidens-democracy-speech.html.

171 "Full Transcript of President Biden's Speech in Philadelphia," *New York Times*, September 1, 2022, www.nytimes.com/2022/09/01/us/politics/biden-speech-transcript.html.

172 Ibid.

173 "Michael Caputo on Facebook Live," *Buffalo News*, September 16, 2020, www.buffalonews.com/multimedia/michael-caputo-on-facebook-live/video_bd8f50c8-f779-11ea-8827-abd983619e35.html

174 Chas Daner, "Trump HHS Aide Michael Caputo Takes Leave of Absence after Facebook Rant," *New York* magazine, September 16, 2020, New York Magazine, www.nymag.com/intelligencer/2020/09/trump-hhs-aide-caputo-attacks-cdc-warns-of-insurrection.html.

175 Noah Weiland, Sheryl Gay Stolberg, and Abby Goodnough, "Political Appointees Meddled in C.D.C.'s 'Holiest of the Holy' Health Report," *New York Times*, September 12, 2020, www.nytimes.com/2020/09/12/us/politics/trump-coronavirus-politics-cdc.html.

176 Ibid.

177 Dan Diamond, "Trump Officials Interfered with CDC Reports on Covid-19," *Politico*, September 12, 2020, www.politico.com/news/2020/09/11/exclusive-trump-officials-interfered-with-cdc-reports-on-covid-19-412809.

178 Ibid.

179 Ibid.

180 Ibid.

181 Reuters Staff, "Fact-Check: 2017 Video Shows Nancy Pelosi Talking about Republicans Using 'Wrap-Up Smear' Tactic," February 5, 2021, www.reuters.com/article/uk-factcheck-pelosi-2017-video-misleadin/fact-check-2017-video-shows-nancy-pelosi-talking-about-republicans-using-wrap-up-smear-tactic-idUSKBN2A52UB.

182 Dan Diamond, "'We Want Them Infected': Trump Appointee Demanded 'Herd Immunity' Strategy, Emails Reveal," *Politico*, December 16, 2020, www.politico.com/news/2020/12/16/trump-appointee-demanded-herd-immunity-strategy-446408.

183 Email from Dr. Paul Alexander to Michael Caputo, Brad Traverse, Caitlin Oakley, Katherine McKeogh, Ruan Murphy, Mark Weber, & Gordon Hensley, July 4, 2020, 11:50 am, www.politico.com/f/?id=00000176-6c7e-d0c3-ab77-7dff0c950001.

184 Email from Dr. Paul Alexander to Michael Caputo, Brad Traverse, Caitlin Oakley, Katherine McKeogh, Ruan Murphy, Mark Weber, & Gordon Hensley, July 4, 2020, 1:44 pm, www.politico.com/f/?id=00000176-6c7e-d0c3-ab77-7dff0c950001.

185 Peter Navarro, "The Immaculate Deception – Six Key Dimensions of Election Irregularities," *Bannon's War Room*, December 15, 2020, www.bannonswarroom.com/wp-content/uploads/2020/12The-Immaculate-Deception-12.15.20.pdf.

186 Ibid.

187 Tristan Justice, "FBI Refuses to Say How Many Informants Were Involved in Jan. 6 Violence," *The Federalist*, January 11, 2022, www.thefederalist.com/2022/01/11/fbi -refuses-to-say-how-many-informants-were-involved-in-jan-6-violence/.

188 Tom Scheck, Geoff Hing, Sabby Robinson & Gracie Stockton, "How Private Money from Facebook's CEO Saved the 2020 Election," *NPR*, December 8, 2020, www.npr.org /2020/12/08/943242106/how-private-money-from-facebooks-ceo-saved-the-2020 -election.

189 David Molloy, "Zuckerberg Tells Rogan FBI Warning Prompted Biden Laptop Story Censorship," *BBC*, August 26, 2020, www.bbc.com/news/world-us-canada-62688532.

190 Post Editorial Board, "Fauci's Direct Line to Zuck proves Facebook COVID Censorship was All About Power and Not Public Health," *New York Post*, September 9, 2022, www .nypost.com/2022/09/09/faucis-direct-line-to-zuck-proves-facebook-covid-censorship -was-all-about-power/.

191 "The Rachel Maddow Show," *MSNBC*, January 3, 2017, www.msnbc.com/transcripts /rachel-maddow-show/2017-01-03-msna973326.

192 "Military Documents about Gain of Function Contradict Fauci Testimony Under Oath," *Project Veritas*, January 10, 2022, www.projectveritas.com/news/military-documents -about-gain-of-function-contradict-fauci-testimony-under/.

193 Ibid.

194 Ibid.

195 Ibid.

196 Andrew Mark Miller, "Fox News Special Report Outlines Fresh Questions on What Fauci, Government Knew About COVID Origin," *Fox News*, January 25, 2022, www.foxnews.com/politics/special-report-outlines-fresh-questions-on-what-fauci -government-knew-about-covid-origin/.

197 Andrew White, "REPORT: Peter Daszak Worked for CIA, EcoHealth Alliance is a 'CIA Front Organization,'" *National File*, January 20, www.nationalfile.com/report -peter-daszak-worked-cia-ecohealth-alliance-cia-front-organization/.

198 Ibid.

Chapter Seven

199 Paul Elias Alexander, "More than 400 Studies on the Failure of Compulsory Covid Interventions (Lockdowns, Restrictions, Closures)," *Brownstone Institute*, November 30, 2021, www.brownstone.org/articles/more-than-400-studies-on-the-failure-of-compulsory -covid-interventions/.

200 Paul Elias Alexander, "The Catastrophic Impact of Covid Forced Societal Lockdowns," *American Institute of Economic Research*, January 30, 2021, www.aier.org/article/the -catastrophic-impact-of-covid-forced-societal-lockdowns/.

201 Paul E. Alexander, "School Closure: A Careful Review of the Evidence," *American Institute for Economic Research*, February 19, 2021, www.aier.org/article/school-closure -a-careful-review-of-the-evidence/.

202 Paul Elias Alexander, "75 Studies and Articles Against COVID-19 School Closures," *Brownstone Institute*, December 24, 2021, www.brownstone.org/articles/75-studies-and -articles-against-covid-19-school-closures/.

[203] Paul Elias Alexander, "More than 150 Comparative Studies and Articles on Mask Ineffectiveness and Harm," *Brownstone Institute*, December 20, 2021, www.brownstone.org /articles/more-than-150-comparative-studies-and-articles-on-mask-ineffectiveness-and -harms/.

[204] Paul Elias Alexander, "More than 400 Studies on the Failure of Compulsory Covid Interventions (Lockdowns, Restrictions, Closures)," *Brownstone Institute*, November 30, 2021, www.brownstone.org/articles/more-than-400-studies-on-the-failure-of-compulsory -covid-interventions/.

[205] Paul Elias Alexander, "150 Plus Research Studies Affirm Naturally Acquired Immunity to Covid-19: Documented, Linked, and Quoted," *Brownstone Institute*, October 17, 2021, www.brownstone.org/articles/79-research-studies-affirm-naturally-acquired-immunity -to-covid-19-documented-linked-and-quoted/.

[206] Carl Heneghan, Jon Brassey, & Tom Jefferson, "CG Report 3: The Impact of Pandemic Restrictions on Childhood Mental Health," *Collateral Global*, (Accessed October 16, 2022), www.collateralglobal.org/article/report-the-impact-of-pandemic-restrictions-on -childhood-mental-health/.

[207] AIER Staff, "Cost of Lockdowns: A Preliminary Report," *American Institute for Economic Research*," November 18, 2020, www.aier.org/article/cost-of-us-lockdowns-a- preliminary-report/.

[208] "Great Barrington Declaration," *Global Declaration*, October 4, 2020, www.gbdeclaration .org/.

[209] John P.A. Ioannidis, "The Infection Fatality Rate of COVID-19 Inferred from Seroprevalence Data," *MedRxiv*, July 14, 2020, www.doi.org/10.1101/2020.05.13.20 101253.

[210] Andrew T. Levin, Gideon Meyerowitz-Katz, Nana Owusu-Boaitey, et al., "Assessing the Age Specificity of Infection Fatality Rates for COVID-19: Systematic Review, Meta-Analysis, and Public Policy Implications," *MedRxiv*, August 28, 2020, www.doi.org/10 .1101/2020.07.23.20160895.

[211] Martin Kulldorff & Jay Bhattacharya, "One of the Lockdowns' Greatest Casualties Could be Science," *The Federalist*, March 18, 2021, www.thefederalist.com/2021/03/18 /one-of-the-lockdowns-greatest-casualties-could-be-science/.

[212] Paul Elias Alexander, "If It's Over, Why the Continued Emergency?" *Brownstone Institute*, September 20, 2022, www.brownstone.org/articles/if-its-over-why-the-continued -emergency/.

[213] Harvey Risch, Jayanta Bhattacharya, & Paul Elias Alexander, "The Emergency Must Be Ended Now," *Brownstone Institute*, January 23, 2022, www.brownstone.org/articles /the-emergency-must-be-ended-now/.

[214] Yafei Liu, Wai Tuck Soh, Jun-Ichi Kishikawa, et al., "An Infectivity-Enhancing Site on the SARS-CoV-2 Spike Protein Targeted by Antibodies, *Cell*, May 24, 2021, DOI: 10.1016/j.cell.2021.05.032, www.pubmed.ncbi.nlm.nih.gov/34139176/

[215] Novara Yahi, Henri Chahinian, & Jacques Fantini, "Infection-Enhancing Anti-SARS-CoV-2 Antibodies Recognize Both the Original Wuhan/D6146 Strain and Delta Variants. A Potential for Mass Vaccination," *Journal of Infection*, August 9, 2021, 83(5), p. 607-635, doi: 10.1016/j.jinf.2021.08.010, www.ncbi.nlm.nih.gov/pmc /articles/PMC8351274/.

216 Paul E. Alexander, "We Must Not Be Forced into Vaccinating Our Children from COVID," *American Institute for Economic Research*, March 31, 2021, www.aier.org/article/why-we-must-not-be-forced-into-vaccinating-our-children-from-covid-beware/.

217 Paul E. Alexander, "Why Are We Vaccinating Children against Covid-19?" *American Institute for Economic Research*, March 20, 2021, www.aier.org/article/why-are-we-vaccinating-children-against-covid-19/.

218 Sivan Gazit, Roei Shlezinger, Galit Perez, et. al., "Comparing SARS-CoV-2 Natural Immunity to Vaccine-Induced Immunity: Reinfections versus Breakthrough Infections, *MedRxiv*, August 25, 2021, www.doi.org/10.1101/2021.08.24.21262415.

219 Paul Elias Alexander, "53 Efficacy Studies that Rebuke Vaccine Mandates," *Brownstone Institute*, October 28, 2021, www.brownstone.org/articles/16-studies-on-vaccine-efficacy/.

220 Scott Morefield, "The Infuriating Habit of Ignoring Natural Immunity," *Brownstone Institute*, September 10, 2021, www.brownstone.org/articles/the-infuriating-habit-of-ignoring-natural-immunity/.

221 David Rosenberg, "Natural Infection vs. Vaccination: Which Gives More Protection?" *Israel National News*, July 13, 2021, www.israelnationalnews.com/news/309762.

222 Paul Elias Alexander, "150 Plus Research Studies Affirm Naturally Acquired Immunity to Covid-19: Documented, Linked, and Quoted," *Brownstone Institute*, October 17, 2021, www.brownstone.org/articles/79-research-studies-affirm-naturally-acquired-immunity-to-covid-19-documented-linked-and-quoted/.

223 Paul E. Alexander, "The Dangers of Masks," *American Institute for Economic Research*, April 9, 2021, www.aier.org/article/the-dangers-of-masks/.

224 Paul Alexander, "Povidone Iodine or Hydrogen Peroxide (diluted) Nasal Oral Wash: This is the Way Out for Nations that Locked Down Too Long and Hard, Waited for a Failed COVID Injection: Not HCQ/IVM, Use Nasal Wash," *Substack*, September 21, 2022, www.palexander.substack.com/p/povidone-iodine-10-or-hydrogen-peroxide.

225 Peter McCullough, Paul E. Alexander, Robin Armstrong, et. al., "Multifaceted Highly Targeted Sequential Multidrug Treatment of Early Ambulatory High-Risk SARS-CoV-2 Infection (COVID-19), *Review of Cardiovascular Medicine*, vol. 4, p. 517-530, December 30, 2021, doi: 10.31083/j.rcm.2020.04.264, www.pubmed.ncbi.nlm.nih.gov/33387997/.

Epilogue

226 Sarah Mervosh, "The Pandemic Erased Two Decades of Progress in Math and Reading," *New York Times*, September 1, 2022, www.nytimes.com/2022/09/01/us/national-test-scores-math-reading-pandemic.html.

227 Lena Sun and Dan Diamond, "CDC, Under Fire, Lays Out Plan to Become More Nimble and Accountable," *Washington Post*, August 17, 2022, www.washingtonpost.com/health/2022/08/17/walensky-revamp-cdc-culture-covid/.

228 Ibid.

229 Jim Hoft, "Flashback: Video Shows Bill Gates Lied to Trump About Dangers of Vaccines and Trashed Robert F. Kennedy, Jr.," *Gateway Pundit*, September 16, 2022, www.thegatewaypundit.com/2022/09/flashback-video-shows-bill-gates-lied-trump-dangers-vaccines-trashed-robert-kennedy-jr/.

AUTHOR'S NOTE BY KENT HECKENLIVELY

A Call from Tony

My personal editor, good friend, and fellow author, Max Swafford, tells me I'm the luckiest writer on the face of the planet.

Not only do I get to publish books I've conceived and written, but my publisher, Tony Lyons of Skyhorse Publishing, often calls me with interesting projects. In my view, Tony is the most fearless publisher in America, and I've told him the motto for Skyhorse should be, "The place where rebels, renegades, and revolutionaries get the chance to tell their side of the story." Tony regularly publishes the works of Roger Stone, President Trump's most enthusiastic backer, as well as that of former Trump attorney Michael Cohen, probably the most definitive anti-Trump book on the market.

It's almost as if Tony believes in the First Amendment.

But that doesn't mean working with Tony is without its challenges. He wants things fast and expects the impossible. And yet more often than not, I've been able to rise to the challenge. At this time in my career, I've written three books about Big Pharma with Dr. Judy Mikovits and Dr. Frank Ruscetti, one book against vaccine mandates (and opposing me was famed Harvard Law professor Alan Dershowitz, who argued in favor of them), one questioning masks as a response to COVID-19 (also with Dr. Mikovits), a book on the childhood vaccine program, one in favor of a treatment called interferon (for which I received a special commendation from *Kirkus Reviews*), and three with Project Veritas whistleblowers, going after Google, Facebook, and CNN.

All told, I've sold more than a quarter million books, have more than ten thousand Amazon customer reviews, and my average ranking is 4.7 out of 5.0 stars.

When I'm working on a project, I'll often hear from Tony every week or so checking in, but when I'm between projects I generally don't have much contact with him.

Which is why when my phone rang in the summer of 2022, and I saw his name on the screen, I answered it by asking, "What kind of trouble are you going to get me into now?"

"The good kind," he replied cheerfully.

"Okay, who is it?"

"Dr. Paul Elias Alexander," Tony answered. "He was the evidence-based medicine expert at Health and Human Services brought in during the COVID-19 crisis."

Tony knows from the books I've written that I have strong feelings about our government's response to COVID-19.

"Okay, tell me more," I responded cautiously.

"He was arguing against the lockdowns, masking, and even the vaccines. Of course, this got him ostracized, even though he was the guy they brought in specifically to make sure their decisions were based on strong evidence. He even talks about what it was like to run afoul of Fauci, who forced him to resign."

"I'm in," I said. From my books with Dr. Judy Mikovits on chronic fatigue syndrome/myalgic encephalomyelitis, as well as my work with thirty-eight-year government scientist and winner of the Distinguished Service Award from the National Institutes of Health, Dr. Frank Ruscetti, it's clear to me that Dr. Anthony Fauci represents one of the greatest threats to public health since the bubonic plague.

"Let me give you Paul's number. He's expecting your call," said Tony.

I called Paul and we talked for about forty-five minutes. From that conversation I was able to come up with an outline of his story, highlighting certain angles he thought were important. Paul is originally from Trinidad and Tobago in the Caribbean, immigrated to Canada as a young man, and has been educated at some of the most prestigious institutions in the world. He is passionate about what he believes, always willing to provide data to back up his opinions, and, in his voice, you can still hear the lovely accent of the Caribbean.

I consider it a great honor to have helped Dr. Paul Elias Alexander bring his story to the public.

Although because of his courage and heroism it is now difficult for him to have a stable income, it is my hope that when scientists and researchers who genuinely value evidence and the public welfare are in power, they will reward Dr. Alexander with a position commensurate with his bravery and brilliance.

Acknowledgments

I must take the opportunity to first thank my direct family for their never ending support, especially during my trying times in Washington, DC. I would also like to thank my mother, Patricia, and my father, Edward, for having faith in me and always encouraging me to reach further. I am fortunate to have the greatest brother in the world, Phillip, and my wonderful three sisters, Faye-Anne, Giselle, and Liesel, and all of their families.

I would like to thank a mentor of mine, Dr. Gordon Guyatt (father of evidence-based medicine), as well as dear friends I have gained on the COVID freedom fighter and COVID expert journey—Dr. Peter McCullough, Dr. Harvey Risch, Dr. Howard Tenenbaum, Dr. Robert Malone, Dr. Naomi Wolf, Dr. Richard Urso, Dr. Ryan Cole, Dr. Pierre Kory, Dr. Paul Marik, Dr. Simone Gold, Dr. Geert Vanden Bossche, Dr. Mike Yeadon, Dr. John Littell, Dr. Vladimir Zev Zelenko, Dr. Parvez Dara, Mr. Erik Sass, Mr. Jeffrey Tucker, Ms. Jenny Beth Martin, and others whom space limitations preclude a full mention. My dear beautiful friends Brent Ford, Warren Ford, Marc Mallet, Yasmin Rezende, Candice Dos Ramos, the Alexander family, and Nicholas family.

Lastly, I'd like to thank my coauthor, Mr. Kent Heckenlively, and my publisher, Tony Lyons and Skyhorse Publishing, for their conviction and faith in me.

—Dr. Paul Elias Alexander

I'd first like to thank my wonderful partner in life, Linda, and our two children, Jacqueline and Ben, for their constant love and support. I'd also like to thank my mother, Josephine, and my father, Jack, for always encouraging me. I have the best brother in the world, Jay, and am appreciative to his wife, Andrea, and their three kids, Anna, John, and Laura.

I've been fortunate to have some of the greatest teachers in the world, Paul Rago, Elizabeth White, Ed Balsdon, Brother Richard Orona, Clinton Bond, Robert Haas, Carol Lashoff, David Alvarez, Giancarlo Trevisan, Bernie Segal, James Frey, Donna Levin, and James Dalessandro.

Thanks to the fantastic friends of my life, John Wible, John Henry, Pete Klenow, Chris Sweeney, Suzanne Golibart, Gina Cioffi Loud, Eric Holm, Susanne Brown, Rick Friedling, Max Swafford, Sherilyn Todd, Rick and Robin Kreutzer, Christie and Joaquim Perreira, and Tricia Mangiapane.

Lastly, I'd like to thank my agent, Johanna Maaghul, the fabulous folks at Skyhorse, and publisher Tony Lyons, for his faith in me shown over the years.

—Kent Heckenlively